LOOK WHAT I STEPPED IN

LOOK WHAT I STEPPED IN

BY PAUL SWAFFAR

ILLUSTRATED BY
TOM PHILLIPS

THE LOWELL PRESS
KANSAS CITY, MISSOURI

FIRST EDITION

Printed in the United States of America
Library of Congress Catalog Card No. 72-93803
ISBN 0-913504-10-6

Designed by Ronald E. Garman

LOOK WHAT I STEPPED IN

DEDICATION

Dedicated with thanks and gratitude to all those in the livestock industry who have worked for the improvement of farming and ranching, to my friends who have shared their experiences and to my wife Xerlan and three daughters who have shared the fun of it all.

CONTENTS

CHAPTER 1

I BEGIN TO NOTICE THINGS

Chapter 1

I Begin to Notice Things

A fellow told me that somebody once made a survey which showed that 73.1 percent of all Americans, with the exception of certain Oklahoma Indians and a few hula dancers, hanker to write a book. The way this fellow explained it, the survey was made by post-graduate students from several non-accredited colleges, years before the computers were even dreamed about. He said that the whole project consumed the better part of 278 days, done mostly by mail and that the surveyors were paid at the hourly rate of 54 cents plus $1.40 per diem for food, hotel and insurance whenever away from their pads. I asked how they could put together any amount of data, sort and shuffle the stuff and come up with a decent answer without computers. He finally admitted that his information was second or third handed but that the way he understood it the analysis work was done by two retired school teachers by the names of Miss Hortense Decinalt and Miss Penelope Phrase.

I belong to the 73.1 percent group and due to the fact that I'm not overworked at anything else, and over strong objections of my former and now senile English professor, I have decided to go ahead.

I told what I thought was a very good friend that I was figuring on writing a book and do you want to know what he said? He said, "What for?"

I told him I thought it would be fun and he said, "Who for?" I said for me and he said he didn't think that was a very legitimate reason and then went on reading the book he was reading. I haven't discussed the matter with him since.

Two of my boyhood idols were Hoot Gibson, the hard ridin' movie cowboy, and Babe Ruth. Hoot never bothered to write a book, as far as I know, but somebody wrote one about baseball and the Babe signed his name to it.

Cowboys hardly ever write books. There are certain exceptions like Charlie Russell, Jack London and Will Rogers, who wrote philosophically and wittily about nearly anything or anybody. Actually, these weren't real cowboys but they fooled around with cowboys enough to know how the ball bounced as they say.

Quite a number of sports celebrities have written books. Some have wound up as best sellers or nearly so depending somewhat on how many "secrets" are revealed.

Since I have not been able to find that anybody by the name of Swaffar has written a book I thought that void in literature should be filled. There was an Englishman by the name of Hannan Swaffar (please note the spelling) who was a widely known literary critic around London, but so far as I know he carefully avoided writing a book, being content to tear to shreds those written by others.

I have never been able to determine just how the name with that spelling "Swaffar" came about. Many's the time I've lain wide awake for hours on Sunday afternoon (before football season) wondering why Swaffar is so difficult to spell. I have been called, addressed and phoned at by these few names amongst many more; Swaffer, Swafford, Swather, Mr. S. Waffer, Swaffle, Swafter, Swanner, Swauffer, Stauffer, and Swatter. There are several others, some of which begin with S but are unprintable and besides I deny them. One lady wrote when I was Secretary of the American Hereford Association as follows:

"Dear Mr. Waffle,

Please write me and return in the enclosed envelope just why I haven't got my registration papers. I have sold the bulls and people think I lied and am mean.

Please hurry!"

My grandpa, Frank Swaffar, once told me when I was out at the barn with him that his grandfather came over to America from Wales. I asked my father, George Swaffar, if he knew anything about it and he said he didn't but that I'd better believe it because he once got a sound threshing for doubting something his dad, Frank Swaffar, had said.

Maybe it would be a good idea to set the record straight right here on this page regarding the purpose of the book in order to avoid an explanation on the last page. I'm not mad at anybody, I'm not crusading for any cause, I have no real story to relate and I have no intention of doing a lot of detailed research in order to quote statistics that nobody would read and I see no object in simply rewriting something that's already been written hundreds of times. I know plenty of dirt but if that's what you're after I suggest "Tropic of Cancer" or "Fanny Hill."

According to my mother who should know, I was born at Lehigh, Oklahoma on November 16, 1907. I mention this date only because of its other historical significance. That was the day on which Oklahoma officially became a state. I therefore have no idea whether I was born in Oklahoma or the Indian Territory, since I have not bothered to find out the exact time of the day statehood became official and my mother says she doesn't recall the exact time either as all she could think of at the time was how utterly ugly I was.

My mother says that when my older sister was born there was no doctor handy so she did what all the other non-affluent women did around there, she had a mid-wife. I remember one time hearing my mother discussing her child-bearing with a neighbor lady. She related her experience with the mid-wife, after which the neighbor asked if the same mid-wife was present when I was born. "Oh, no," said my mother, "we just had a doctor with him."

My first recollection of learning anything about animal husbandry is that of being kicked in the mouth by a mule colt. This happened when I was around four years old and most of my relatives and certain other enemies stoutly maintain that I've been addled ever since. This experience, however, did teach me a very important lesson about mules. Never, never, never loiter or even slow down behind a mule, no matter how innocent it may appear. Although mules have almost entirely gone out of style, you may encounter one sometime in which case it would be well to approach same from the front end or else not at all. I might mention in passing that I have learned several other things about mules, none of which seems to justify a full discussion of the subject.

I cannot, however, pass up the opportunity of telling about one old mule we had. His name was Coalie. I suppose that name came about by virtue of his being pitch black. He was mean, stubborn, cantankerous and very intelligent. We had a neighbor who borrowed him one time to plant his oats when one of his own had "Monday morning colic." He told my father that Coalie was "the hatefullest goddamn mule I ever looked at the ass end of."

In those days we cultivated the corn with two mules hitched to a walking cultivator. One was Coalie. About 10:30 A.M. when we would reach the row end nearest the barn, Coalie would be panting heavily, head and ears drooping, with an expression of near exhaustion. However, when the opposite end had been reached, and we turned again toward the barn, Coalie would be stepping lively as a colt. He had a mischievious habit of clamping down tightly on a guide line when it got underneath his tail which often happened at the turns. Dislodging the line was a difficult and irritating task for an average sized boy which I was locally considered to be. We had a neighbor kid who was slightly prankish and slightly daffy. One day when he was at our house, my mother was baking sweet potatoes and when they were "done" and piping hot she gave this kid and me a couple to eat, one of which was long, small and thin. We went out to the barn where Coalie was stabled and the neighbor kid climbed up on the side of the stall, reached over and placed the hot potato under Coalie's tail. Of course, Coalie clamped down thinking it was the usual line. Have you ever seen a mule have a hard fit? Well he had one. He literally tore the stable down but he was too stubborn to let go until he had scorched his rear. Thereafter he was relatively sure it was the line before he clamped down. In case you are not familiar with the history of the expression "stubborn as a mule" you weren't raised on a mule farm.

At the time I was born my father was a top-side hoist operator at the local coal mine. The mine soon thereafter folded so dad took up part-time farming and part-time Baptist preaching around a place called Standing Rock, Oklahoma. He had very little formal education, having gone only through the fourth grade, but somehow learned to read exceptionally well for a fourth grader. He read very little except the Bible. This book he knew almost in its entirety from memory but very few days passed in which he failed to read it some more. In case you have never had experience around a Baptist preacher, I can tell you they are some kind of strict. About the only things we kids were allowed to do out in the open was to attend the 9:00 A.M. to 11:00 A.M. Sunday School, 11:00 A.M. to 1:00 P.M. morning preaching, 5:00 P.M. to 7:00 P.M. B.Y.P.U., and then 7:00 P.M. to 9:00 P.M. Wednesday night Prayer Meeting. Of course we did a lot of things that weren't allowed.

Soon after my encounter with the mule colt, we moved to Boggy Depot, Oklahoma, which was a stage coach and cattle resting place in the early days along the Shawnee Trail. So far as I know, Boggy Depot has long since been stricken from the map. Along about this time I began to notice things, and one of the things I noticed was some of the old-time cattle, a few of which still moseyed through that country on their way from Texas to the Osage grass country or other places where there was grass. Most of the Trail Drives, even the short ones, had ended around 1900.

Would you like to know what these animals looked like?

Well, they didn't look like very much. You'd see more colors than are in the ordinary rainbow, but as I recall there were more brindles than anything else. In case you've never seen a brindle maybe I'd better describe one. Brindle is about the color of an old straw stack that's been splashed with black mud by a passing truck. Most of those who have studied the brindle are agreed that this color comes from a mixture of Jersey, Longhorn, (whatever that is) maybe some Spanish fighting bull, once in a while some English blood—but not often—and with a lot of just plain cow. In addition to the brindles, there were some solid reds, yellows, spots, buckskins, roans, blacks, greys, light blues, tans, a few whites, and various mixtures of these.

I don't recall seeing a green or a purple but I wouldn't swear there weren't some.

I figure that if some of these crossbreeding advocates have their way we're apt to return to these multi-colored critters.

One thing about these "trail cattle," they were fairly uniform in conformation—sad-faced, long necked, extremely long legged, mule rumped, and with bodies as long as from here to way over yonder. But could they ever travel! No doubt there have been more than a fair number of lies told about how far some of these drives went in a day and I remember my Uncle Lee telling about one that he was on which covered fifty miles. Today, about the only thing that resembles an old time cattle drive is in Gunsmoke, a western movie, or when cattle in the mountain sections are being moved from winter quarters to high summer pasture or vice versa if there is no passable road for trucks.

We didn't live at Boggy Depot very long as the land wasn't very rich and part-time preaching alone hardly paid enough to raise a family of kids which later became seven in number. We moved to a fairly sorry farm near Atoka, Oklahoma, and my father kept right on part-time farming and part-time preaching, managing somehow to eke out a living until around World War I when, because of a little political pull, he landed a job at Warner, Oklahoma, managing the farm of the Junior College, and where we older kids, and more to come, could attend school which Mamma saw to it we did. God bless her, she worked eighteen hours each and every day and beat us on the south end if we even looked like we didn't want to go to school.

My next recollection of animal husbandry is that of learning to milk the family cow and slopping the pigs. Learning to milk is quite an experience for a young lad of around eight years. At our place we had no such thing as a milk barn or even a shed. The cows were milked in the corner of a lot, summer and winter alike. It was the hottest farm chore of all in the summer and the coldest in winter. In the summer the cows received absolutely nothing to eat with the exception of what they gleaned from the rather poor pastures. Early in the spring wild onions were abundant and later on in the summer these were replaced by bitterweed. Consequently, the milk and the homemade butter most always tasted like either onions or quinine. In case you're not acquainted with quinine, it's a white powder, so-called medicine, used internally for treatment of malaria, a rather common ailment around Atoka and Warner, Oklahoma, where there were hoards of mosquitoes in those days. Our cows' winter rations consisted of corn "nubbins," corn fodder, and if we could afford it once in a while some cottonseed cake. The cows gave very little milk in the winter, or summer either for that matter. There was one advantage in this. It didn't take long to get the job over with. Flies and manure were a real problem in the summer-time. Various things attracted flies to the milking spot and rarely ever did you fail to get at least three or four in the bucket during the milking procedure. When you did, you simply flicked them out with your forefinger and told nobody about it. The cow, in her attempt to swish away the flies with her tail would often manage to drop a few blobs of manure in the bucket or hit your face or your shirt. If my mother noted that some had gotten in the milk, it was fed to the pigs and we got no milk to drink the following day. If she found none we drank the milk although it often did taste a little funny. Refrigeration of any kind was unknown at our house so the milk and butter were chilled by lowering them into the dug well with a rope. Occasionally some spilled over. This didn't do too much for the water's flavor.

At the Warner farm which my father managed, there was quite a good sized herd of Jersey cows which I received 15¢ an hour to help milk. This herd held the state record during that time and I believe the record still stands. They were the hardest-to-milk cows in the whole state.

My next animal husbandry experience came about when my father and the county agent conferred and between them decided I should become a 4H Club member. Several projects were discussed but I was not in on the discussion. Later on, Dad said that among the projects discussed were chickens, vegetable gardening, row crops, sheep, cooking and canning, pecan tree planting, rug weaving, debating, baby beef and pigs. They decided I'd probably fit pigs best, so with a little help from our bank and some rather loose credit by a local red hog breeder, I was able to get a Duroc gilt with this hog breeder holding full title until she was paid for, theoretically with proceeds from the sale of the first litter. I might mention that later on it leaked out that this breeder let the Club boys have the gilts at the "reasonable" price because he couldn't sell them anywhere else.

Anyhow, the first litter came right on time—ten pigs. The clumsy mother laid smack down on the whole bunch, killing four, so I was left with six. I fed, watered, brushed and babied these for approximately eight months, always smelling like a pig pen at meal-time or at school. All girls, including my own sisters, carefully avoided getting anywhere near me especially when I was next to a pot bellied stove. Anyway, when the six pigs were around eight months old, they were sold along with other Club members' pigs at the County Seat Auction. After paying for the mother pig and the feed bills I had \$1.13 left but I still had the sow, now heavy with pigs again. Unfortunately she wasn't as heavy as she looked, giving birth to two lousy pigs. My father reasoned that such a poor litter would not happen again and we'd best give her another chance. This we did. The third litter—another two pigs.

That winter we ate the sow and I went out of the hog business not returning to same until about 1940 when another fellow and I tried garbage feeding, which I may tell you about later on. In case I don't get around to it, I can relate that this garbage feeding venture was also a huge bust and confirmed what I should have known all along—I simply wasn't and ain't no hog man.

Somehow I managed to finish the local high school requirements and was then confronted with the same problem most high school graduates face. What now? Going wages around our country averaged about 12½¢ an hour at the time—1924. The better jobs paid

$2.00 a day but you were expected to put in 16 hours to get it. In addition, there were hardly any jobs to be had.

I thought about several vocations, among them chicken farming, bus driving, railroad engineering, soda-jerking, deputy sheriffing, and going to South America. My mother said that she thought I ought to get more schooling as I didn't appear to know much about anything and that I'd probably have a pretty difficult time to keep from starving unless I learned more about something. My father agreed, as he usually did, so with two pairs of pants, two shirts, a change of underwear and $34.00, I took off for college. Mind you, I was considered fortunate as some of the local boys did not have a change of underwear nor a thin dime in cash—in fact, some had no underwear at all.

I had sort of always had a hankering to know more about agriculture, so on the way to college I decided to try that, especially about animals. I have never regretted the decision. Somehow by picking up odd jobs, and with a few dollars from home now and then, I managed to get through the first two years, "batchin'" most of the time with three other "pore" boys. We lived mostly on bologna, beans, sow belly and anything else which was cheap and filling.

After two years I decided I'd had enough of these meager rations. I figured it would be nice to get a job and make enough money to live in style the last two years. By some maneuvering I landed a job teaching a one-room country school for which I had absolutely no qualifications. I was without doubt the poorest teacher our county ever had, the County School Superintendent told me so. I could write a large book about this experience. I shall, however, limit the report to the following short statement: I learned more than the kids, concluded that one-room schools should be abolished forever, saved not a nickel in two years, and wound up going back to college with less cash than I started with the first time.

The last two years of college were real fine years. I worked at the animal husbandry barns, several other jobs, borrowed considerably from relatives, friends, student loan funds, managed to make reasonable grades, to be fairly successful at livestock judging and having a big lot of fun to boot. I went out for football, lasting until the first scrimmage session.

After graduating from Oklahoma State University (then known

as Oklahoma A. & M.), by a little luck from my livestock judging, I was invited to take a teaching fellowship at Virginia Polytechnic Institute under Professor R. E. Hunt. Professor Hunt was a fine man, treating me with patience and much more respect than I deserved. As I look back now, it's hard to understand his tolerance. Somchow I managed to eke out a Masters Degree there. I thought some about trying for a Ph.D. However, for numerous reasons I gave up the idea, the principal ones were (1) my scholastic record didn't exactly cause people to stampede to get me, (2) I was married to the same wife I've always had, we had two children and no money and, (3) I decided a Ph.D. degree would be absolutely wasted on me as it is with certain others who have only that to recommend them.

Since that time, mostly because of luck and by virtue of being at the right place at the right time, I have been Animal Husbandry Specialist at Virginia, and North Carolina, Manager of the Atlantic Rural Exposition in Richmond, Virginia; then Fieldman, Field Service Director, Assistant Secretary and Secretary of the American Hereford Association. In these jobs I've had the good fortune to travel in every state in America except Alaska, and in a good many foreign countries.

Now that I'm retired from the American Hereford Association I am realizing a dream of living on my own farm. Later on I plan to do a chapter entitled "What Not To Do in Starting a Farm." I've made every mistake the law allows, but so far I've had more fun than anybody who ever tried it.

CHAPTER 2

BEEF

Chapter 2

Beef

My dictionary defines beef as follows: (Plural-beeves, or beefs) 1. The flesh of a slaughtered adult or near adult bovine animal. 2. Any adult or near adult bovine animal, as an ox, cow, steer, bull, etc., fattened for the table. 3. (informal) human flesh; muscle, brawn. 4. U. S. slang, a complaint-v.i. U. S. slang to complain-to beef up (informal) to strengthen or reinforce.

Under the definitions there are several things that should be called to your attention. First, beef can be a food, either dead or alive, it can be a 290-pound football tackle, it can be an ordinary bitch about something in the government and it can be the process of building up the strength for something—like taking vitamins and tranquilizers before facing an irate boss or wife.

Most cowboys and similar people know all about bovine beef but there are many non-cowboys like those living on Manhattan Island or Cape Cod who need to be up-dated on the subject. Notice that item 2 of the definition says "adult or near adult." There's hardly any doubt about what constitutes an adult as most everybody I know will agree that adult means full grown. I may get some hot letters about this, but I figure a full grown bovine would be around four years old. The critical question is then, when is a beast a near adult? It might be reasonable to conclude that a three-year-old would be a near adult, a two-year-old questionable, but a yearling would be something else.

I have reviewed all the literature on the subject and I find there is a substance called "veal," another called "calf meat," and still an-

other called "baby beef." Most Ph.D's on the subject agree that veal comes from very young bovines whose mothers nine times out of ten are milk cows and that it's a damn shame to kill these innocent little fellows since the resulting meat tastes slick, without flavor and UGH! Calf meat is slightly better but you've got to be fairly hungry to think so—maybe stoned. Baby beef—it's a sight how this term is bandied about in high class restaurants, non-high class restaurants and ordinary beaneries on poorly lit side streets. I swear to you that during my forty years of traveling I've eaten the following items at different places, all called "baby beef"; almost white meat from day-old calves; ground chuck with corn flakes added; flank steak meat from Jersey cows; tough round that had been pounded with a claw hammer; paper thin, so-called steaks from old bulls, and others too numerous to discuss in this thesis. Lest you conclude that all baby beef will fall into one of the above groups, I'd better add that I've had as fine a strip sirloin as you've ever read about called baby beef. If I ever run for the Senate and get elected I fully intend to introduce a bill which will require all eating places to properly identify baby beef. Failure to do so will carry the penalty of a lifetime diet of veal. Some people seem to like veal, a fact that I am unable to explain.

Next, observe that the definition says, "a steer or cow fattened for the table." The important word in this part of the definition is FATTENED. In some high (and low) places fat has become a dirty word. I haven't been able to understand exactly why, but a good many medical doctors are all gung-ho about advising people not to go anywhere near a little beef fat. They say it will clog up your blood vessels and that you're apt to go to your Maker long before you're due. There is considerable evidence that the amount of fat found in a decent piece of choice beef is not enough to cause early heart trouble, falling hair, double chins or anything else normally found in the senior citizens' group. The non-animal fat doctors have an ally in a certain group of cattle promoters. This group tries desperately to con all cattlemen into growing animals that will absolutely lay on no fat at all even if fully confined in an Iowa corncrib all winter long. The trouble here is, that kind of an animal will not fetch top prices from the butchers and the completely non-fat beef tastes lousy. It is tough like jerky and it is best suited to stuff peppers. It also makes quite nice dog or cat food. I understand that some of the "cutting" or "breaking" houses are equipped with Homelite Chain Saws to cut it.

Something should be said about beef as associated with "muscular flesh or brawn." The best example of brawn I can think of right off is Buck Buchanan, defensive right tackle of the Kansas City Chief's football team. Buck stands tall at around 6'8" barefooted, he weighs about 290 pounds completely dry and everybody who's felt him, including several quarterbacks, says he's as hard as a terrazzo floor. So much for football and other forms of brawn.

The "complaint" connotation for beef suggests the following synonyms; constructive criticism, griping, yelping, and several others the most common of which is "bitching."

As previously implied, beef as a food can be almost any color, ranging from almost white to almost black. Neither extreme will excite very many people and pinkish red is the best color, I don't care what anybody says.

Beef is the favorite food of most Americans. During 1971 each man, woman, widow, widower, divorcee, teenager and kid averaged eating somewhere around 112 pounds of beef. Most young kids will not eat that much, but traveling men on expense accounts will take up the slack. I traveled for about forty years so I ought to know. 112 pounds times 210 million people makes about 23½ billion pounds, equal to around 40 million steers weighing 1000 pounds each.

There are numerous reasons why beef is a preferred food, such as (1) it tastes exceptionally good when properly prepared; (2) it contains large amounts of protein, calorics, minerals, vitamins and other goodies; (3) it will make armies fight harder, ditch diggers dig deeper ditches and harvest hands pitch more wheat. It will also make white collar workers think more and spend less time around the water cooler.

There are several different kinds and grades of beef which I shall attempt to classify and analyze for the uninformed. The informed can skip to the next chapter.

Raw Beef—Most people don't seem to care too much for raw beef, although it is often fed to football teams and wrestlers. R. J. Kinzer, long time secretary of the American Hereford Association, used to take raw ground beef, mix it up with hot pepper, salt, tabasco, onions, horse-radish and hot stuff from several other bottles and then, without cooking, eat great gobs of same just before devouring a huge medium rare steak. Of course, R. J. was one of a kind, the mold having been destroyed after he was finished.

Usually raw beef is used for zoo animals, fox-hunt packs and middle line backers. I cannot, without considerable reservation, recommend raw beef to plain every day people since I have noticed a good many will gag when they first try it—like with the first go at raw oysters.

On the other hand, cooking any kind of beef until it looks and tastes like an old cow chip should be no less than a misdemeanor, maybe even a felony. Which reminds me of the time Johnny Jones, Leon Freeze, Bob Day and I were traveling by car for two weeks in Old Mexico. We took along an icebox filled with prime KC Strip Steaks from Eddie Williams' Steak House in Kansas City. We kept them cool in the trunk of our car. I can't right off recall just why we took them along as in most any good-sized town in Mexico you can get quite good beef. We would stop most nights in one of the larger towns, hunt up a good cafe and bribe the cook into cooking our steaks. The bribe usually consisted of several pesos or a quart jug of tequila. Mexican cooks are excellent, so we lived high up on the hog. On at least two days we were caught out in the country and each time we could find no place to eat lunch. In addition to the steaks in our icebox we had tomatoes, hot green Mexican peppers, potatoes, loaves of bread and a bottle of sotol juice. It was mesquite brush country but with very little of that. Johnny and Leon always got hungry promptly at 12 Noon and demanded to eat wherever we happened to be. On these two days we therefore stopped promptly at 12 Noon in the country where there was absolutely no wood with which to build a fire for cooking. You've already probably guessed what we did. Johnny simply gathered up a pile of dry cow chips and built a good hot fire. We all had a brief go at the jug, then baked potatoes, peeled tomatoes, cooked the steaks rare and had as fine a meal as anyone could wish for.

Beef Steak–That great mass of humanity known as the "consumer" prefers steaks to any other part of the animal's body as evidenced by the fact that they're always willing to pay more per pound than for any other such as neck, lower flank or right hind shank. Steaks are of several kinds–there are T-bones, porterhouse, sirloin, round, tough, tender, thin, thick, juicy, dry, large, medium, small, minute, mi nute', New York Cut, K.C. Strip, baby beef, hamburger, Swiss, and Lord knows what else.

By chance you haven't heard, there is no finer food on earth than a good choice or prime steak, properly marbled, cooked medium rare and served by a well framed mini-skirted waitress. The only one who would disagree with that would be a jealous wife sitting at home with a house full of fighting kids.

The best steaks come from the loin of a heavy-muscled, well-fattened steer weighing around 1000-1200 pounds on foot and yielding a carcass of about 600 plus pounds. It is improper, disgraceful, and should be unlawful to cook and eat a steak before it is properly aged. For the benefit of the uninformed, aging of beef is somewhat different from aging of people. As people age, they usually get crankier, tougher, and more dried up. Beef on the other hand gets more tender, better tasting with far better flavor. I shall not attempt a detailed discussion of beef aging except to say it consists of hanging the freshly slaughtered carcass in a cooler at approximately 36 degrees temperature for around three weeks time–more or less, depending on the preference of a particular customer. Anyone who cuts a good steak thin should be exported. The proper thickness of a loin steak is from 1½ to 2 inches—some even prefer them thicker. Then, as previously stated, if you cook this just medium rare, you've really got something worth living for.

I got the Better Homes and Garden Cookbook down off the shelf, turned to page 24 and confirmed what I had suspected all along—there are more ways than one to ruin a good steak. There are probably thirty or forty million people in the U.S.A. who will disagree, but I have never seen much sense in paying three dollars for a prime strip sirloin then taking it out in the back yard among the flies, ants, dogs, neighbor kids, overhead birds, whirlwinds and other kinds of pollution, then burning it over a batch of charcoal that cost $1.98. My wife's always nagging me about this. She claims I'm eccentric, lazy, shiftless and anti-social. Some years ago, I bought her an electric stove with two ovens, our electric bills are almost always paid and all she has to do is set that little gadget at whatever temperature the book says, punch a switch and we can both watch television while the meat cooks. Charcoal is all right in its place like heating branding irons and in a big place like the Golden Ox, but in MY back yard—phooey!

I notice also that steaks may be pan-broiled, chicken-fried, home-styled, minuted, skillet-pizzaed, oven rounded and Swissed. During my near half-century of eating out, I've had them fixed all these ways and I will not at this time knock any of them. Just in case you are not aware of it, however, no method of preparing and cooking has yet been discovered that will make an ornery piece of meat into a pleasing steak. This is a fact and it bothers me quite a bit.

Roast Beef — Except for the bones and viscera, about any part of a beef carcass may be roasted with success ranging from excellent down to very poor. Starting with the very poor we'd be obliged to include the nape of the neck, lower flanks and the dewlap. You say you never heard of dewlap? The location roughly corresponds to the upper chest region of your girl friend when she's doing push-ups. There the similarity ceases. Meat from the animal's aforementioned parts should be ground and used for hamburgers, bologna or other forms of sausage.

I'm not quite sure whether rump or chuck roast should be ranked higher since I've never been able to distinguish between the two by taste. I'm aware only when advised by a printed sign or by being told by some smart aleck.

The tastiest and costliest roasts come from the rib. A standing rib looks something like a tiara or Prince Charles hat. You pay for a considerable amount of bone when you buy one but lordy, lordy they're good. The tastiest part is around the rib bones, but getting to it in a public eating place poses a ticklish etiquette problem. The best way to solve this is to bring along your doggie bag, or minus one you can ask the waiter for one. Pretend you are taking them home for your poodle. When you get them home simply ignore the dog, if you have one, and go out on the patio or in the garage and gnaw on the bones. If any of the neighbors happen to catch you, you can say you are following your dentist's tooth exercise recommendations. Prime rib is also the term often used in describing the rib roast. I've always been a bit leery of most "prime rib" entrees on the average eating house menus—they seldom are "prime."

The housewife may like the rolled rib roast which has the bones removed. The "king" of roasts is the Rib-Eye Roast from a good piece of beef. It is new in some markets and you may have to ask for it from your meatman. It is easy to carve.

You can stir up a rousing argument in most any crowd regarding the relative merit of steaks and roasts. Most men—excluding the carving show-off—will prefer steaks. It isn't because the steaks always taste better—it's that carving business. For instance, I've been practicing for some forty odd years and I still can't cut two pieces the same thickness. I get gravy and little slivers of meat all over the good lace tablecloth. I often cut a finger and I hardly ever get by without going to the floor mat a time or two with the thing before I finish. All this with a prime cut. I wind up a mess and my wife never fails to quip to the guests, "You'll have to excuse him, he's only been doing this since 1930."

Like steaks, roasts without any inner muscle fat (marbling) cause a great many divorces. They taste something like rubber roofing without seasoning.

Hamburger — In spite of the misleading name, hamburgers are made from ground beef with or without additives. When additives are used, they can be about anything—soybean meal, old bread crumbs, stale crackers, hominy grits, ground corn cobs or anything lying around handy. As for me, a hamburger should contain only beef. The dictionary says, (1) Ground or chopped beef. (2) Such meat fried, broiled or baked in the form of a patty: also called burger, Salisbury steak, hamburger steak. (3) A sandwich consisting of such meat placed between the halves of a round roll. (named after Hamburg, Germany).

Actually, a good hamburger is a cuisine delight. A really good one is made with freshly ground tender beef properly seasoned with seasoned salt and pepper and garnished with pickles, a small amount of lettuce or tomato (perhaps onion) and with mustard or catsup added or left off according to taste. Place on a grilled bun half! A grilled bun will not become soggy from juices in the meat. Usually, previously ground hamburger meat which you buy at certain stores is hardly fit to eat being made of scraps, tough parts of tough carcasses with whatever else added that the butcher can find no use for. The immense popularity of the hamburger has been brought about by the kids. This is a phenomenon difficult to explain. Actually I've never heard a satisfactory explanation for kids' behavior, anyhow, although I've read considerable about it by writers who had no kids of their own. It is a great thing, however, that kids are so gone on hamburger, otherwise, what on earth would we do with all the old used bulls and cull cows?

Beef Concoctions – It would probably take several pages to name all the dishes which start with some kind of beef and wind up with most anything you can think of mixed in. Take stew for instance. A really good beef stew is something worth waiting up for, but unless you're aware ahead of time just what all went in it your chances of getting poisoned are excellent. When prepared in your own kitchen by other than an upset wife or other cook, your chances are real good of having something tasteful. Get the cook riled up, however, then beware–she can slip in about anything without your detecting it until it's too late as color, texture, flavor, consistency, weight and ingredients are never standard. At public eating establishments beef stew may or may not have any of the following ingredients: Beef, potatoes, pork, celery, rice, carrots, onions, garlic, alfalfa, tomatoes, cull prunes, rutabagas, okra, salt, saltpeter, pickled herring, penicillin, wheat, rye, bourbon, gizzards, oatmeal, buckeyes, field corn, veal, tripe, chitlins, mutton, mountain oysters, and several others I will not have room to list. Normally, I am a gambler enough to order beef stew wherever and whenever I see it on the menu, but the older I get and the more poisonings I have, the less I'm apt to say, "Gimme the blue plate with your stew." Lately companies have been canning the stuff which should be unlawful.

Then there is beef hash—you can find this on breakfast, brunch, lunch, snack, supper, dinner and TV menus. If I were you I'd be mighty careful about ordering it. At breakfast it's usually "Beef Hash with Eggs." I have no proof of this, but I'd venture that Jimmy the Greek will advise that the odds are 12-1 that the beef from which breakfast hash is made came from a tough roast no one would buy the night before. When it's washing or housecleaning day at home, especially on a Monday, hash is apt to appear on the table. I think that left-over beef should be fed to your dog. Some dogs refuse.

A very good concoction made with little squares of beef is "kabobs." I suppose these can be made from most any part of the carcass but round or sirloin is preferred. The beef is whacked into cubes say 1½ x 1½ inches, marinated and then pierced on a long slender skewer alternately with mushrooms, cucumbers, tomatoes—sometimes onions. I'm not aware of how they are cooked, because I might have to do it, but when properly done and piping hot you've got something.

In addition, there are stroganoffs, corned beef, meat loaves, beef liver, chipped beef, various beef sausages, and so on ad infinitum.

Many a housewife, tired, and with a hungry family, has been rescued at meal-time by taking good ground beef, mixing in a little cream, seasoned salt, Worchestershire sauce, forming it into patties about one inch thick, rolling it in flour, and frying it to golden brown on both sides.

Beef is a very handy meat.

EVERYBODY WANTS LEAN, TENDER BEEF THEY ALSO WANT IT TO TASTE GOOD

Present-day meat merchandising methods and housewife demands have placed the cattle breeder in something of a quandary. He is faced with the difficult choice between maintaining the early maturing type we've learned to associate with very high quality meat or of sacrificing some eating quality in favor of more volume and a greater proportion of muscle (red meat).

It is important first to understand just how these two factors—merchandising methods and housewife preference—have affected our business. Not too many years ago, when our lady of the house went to the "butcher" shop, the shop owner would bring out a whole side or quarter of beef, lay it on his block, and cut off her purchase right before her eyes. Generally you could pretty well tell from what kind of critter your beef came. If there was a trifle more fat than our lady wanted, it created no special problem as the butcher would trim fat down to what she wanted, throwing the remainder in the scrap can.

Old-Time "Butcher" Shop Now Gone

This is all changed now. There is hardly a "butcher" shop left. Instead, in most food stores, there is a long meat counter, generally extending across one side of the store, which displays everything from fish through fowl, pork, sausages of all kinds and every cut of beef imaginable. Beef cuts are sized, labeled, graded and packaged in cellophane. Unless you are an excellent judge or read the grade label—if one is available—you haven't the slightest notion what kind of animal the cut came from. You buy by looks only or by the reputation of the beef customarily handled by the particular store.

Except in rare instances, or in the case of experts, our lady tries to buy beef with the most "red meat," hoping it will be tender enough to suit her family. One thing for sure, however, is that she will not buy tough meat if she knows it. If she is among the more acknowledgeable—and there are many in this class—she's apt to select a cut with slightly more fat—marbling—because she knows it will be better flavored.

Most of us believe that quality beef has been responsible for a major share of present-day beef demand. It is this, we think, that has created a national beef appetite of somewhat more than 110 pounds per person per year.

Question Asked and Answered

Just what is quality beef anyhow? Someone somewhere has described quality as a combination of the palatability of the edible portion of a carcass and its physical composition. Palatability is determined by a combination of tenderness, juiciness and flavor. Tenderness usually is regulated by age—young animals being tender and older ones tough (less tender). Flavor is regulated by a combination of the intrinsic substances in the muscle itself and the intra-muscular fat, known as marbling. Juiciness is generally a function of marbling —the more marbling the juicier the meat.

So in the three basic qualifications of quality meat—Tenderness, Flavor and Juiciness—one is regulated primarily by age while the other two are in the main governed by proper marbling. Since very young beef is without doubt the most tender, why not slaughter all cattle at a very young age, thereby guaranteeing tenderness? There are two reasons why not: (1) Very young beef is tasteless, anemic looking and generally unprofitable for a grower and, (2) We have to find use for animals whose productive life is gone or which have become inefficient.

Throughout the years, breeders, feeders and processors have attempted to breed, feed and merchandise a beef product that is young enough to be reasonably tender but old enough to be tasty and palatable—all the while trying to do so at profit enough to live with and by using feeds, for the most part, that can only be utilized by beef cattle. It hasn't been easy.

For some time now we have been told by some that housewives demand all lean meat and that the industry had better get along with the job, or else Mrs. H. W. will go completely to poultry, fish and beans. Some of this is true and we may as well face it.

So the cattle breeder is confronted with a most difficult problem. Just how many of the long-sought elements necessary to make this tender, flavorful, juicy beef can he afford to ignore—or change—in favor of more bulk, less fat and a leaner carcass. Some changes may be in order—but how much and how fast is the question? Cattle breeders can ill afford to make a costly mistake. We need to know where we're going and to be sure that a change is "right" for the entire industry. We must always remember that functional soundness and efficient reproduction also are vital requirements.

Considerable Risk is Involved

No doubt by altering somewhat our beef animal's conformation and by some changes in nutritional procedures we can produce a huge, all lean animal of near the proper weight. The real question is: Can we afford to do so at the risk of losing some of our beef demand? Should we make changes if production efficiency as a result, will be sacrificed?

For whatever it is worth, it is my opinion that we must ferret out within our breed those animals which will adhere to our long and successfully established measures of good conformation, but which will also gain weight somewhat more rapidly and make more efficient use of available feeds, while at the same time laying on—and in—enough fat to make the muscle acceptable from the standpoint of palatability. We do have such cattle, and we don't have to adopt sudden and radical changes in their general conformation to accomplish this job. In fact, we will probably find more and more that those animals which are best suited for the future will fall within those groups with the most desirable modern conformation.

In spite of much that has been written and said about the need for change, our cattle are efficient. I do not mean they can't and shouldn't be improved. Many feed tests and feeding operations are getting three pounds or more per day gain in the feedlots, much of which is done with roughages. And roughages still supply the main ingredient of economical rations.

We can and must make every reasonable attempt to satisfy the beef consumers. At the same time, it is extremely doubtful whether we can afford to sacrifice all of the things so necessary to make palatable, tasty beef in order to get a no-fat, tough and far less tasty product. To do so will guarantee that Mrs. Housewife will change to some other source of protein.

CHAPTER 3

FARMING IS FUN

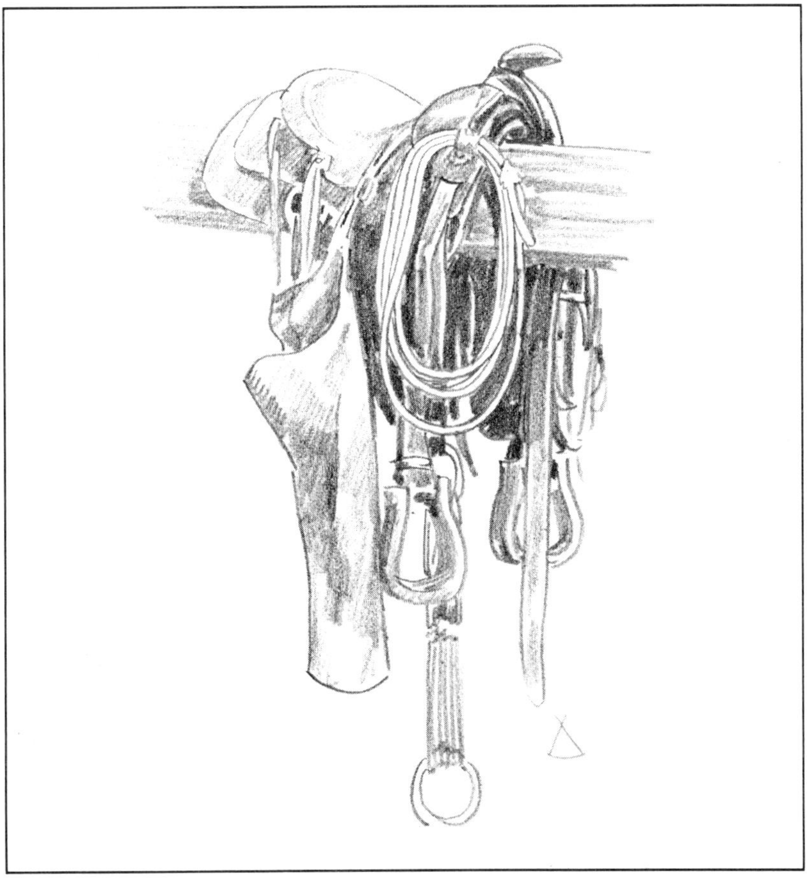

Chapter 3

Farming is Fun

In bygone days if you decided to farm here in America there were several ways you could go about getting your hands on the necessary land. Far enough back you might get a GRANT if you stood in line and your politics were right. This method has been obsolete for so long it hardly seems worthwhile to discuss it further. If you're interested in that facet of history let me know and I'll list several books in which you can bone up on the matter.

I would not wish to go on record urging anyone to commence farming or ranching, but if you're determined to do so and can find no other way of expending your energy, I can guarantee you'll work harder, get tireder and make less money than at most anything else you might attempt. But you'll have a barrel of fun, you'll be healthy, happy and probably live longer than you expected.

In addition to grants, there are five ways by which ownership of a farm or ranch has come about. They are: (1) inherit, (2) marry, (3) homestead, (4) steal, (5) buy. Of these, one is no longer possible, one is illegal and two are limited.

After careful consideration and after consulting numerous economists, star gazers and ordinary cowpokes, I have concluded that inheriting is the very best way although this method isn't as easy as it once was. The main reason it isn't easy is because the tax people (IRS for short) have decided that you really have no business inheriting a place without paying through the nose to do so, even though your father, grandfather, and great-grandfather worked their daylights out to put the place together so you could have it when they all went to the last round-up.

Some of the best spreads with the most efficient management and the greatest contribution to our agricultural economy, red meat production and real stability are the outfits passed down from father to son.

I do not wish to discourage you from inheriting a place, but before you do you ought to know that you don't just inherit one day and start right in accumulating a fortune the next day. If there's someone from whom you expect to inherit a spread and that person is still living I suggest you talk it over and do your best to get him or her to give it to you piece by piece so you won't have the entire tax lug when they pass on. It might be a good idea to have a real brilliant lawyer hanging around during the talks. One difficulty you may run into is that the one from whom you expect to inherit, nine times out of ten, expects to outlive you.

Possibly an inheritance tax is fair if the tax law is properly interpreted and fairly administered. Many revenue agents take the position that fair evaluation of land is whatever other land in the approximate locality is selling for, regardless of the size of the parcel or regardless of the land's earning capacity. Stephen Hart, legal counsel for the National Livestock Tax Committee and who, with his associates has studied and argued agricultural tax matters more than anyone in America, has this to say on the subject:

"The prime reason for the upward trend in valuation is the Internal Revenue Service's interpretation of the requirement in the Federal estate tax regulations that the estate tax be imposed on the market value of the property held by the decedent at the time of his death. Today the price for which farm or grazing land might sell to speculators is out of all proportion to what it will earn for farm or grazing purposes. Unfortunately, however, many revenue agents refuse to give any consideration whatsoever to the earning capacity of a ranch or farm in determining its value for estate tax purposes. To the contrary, they rely on the inflated sales prices of similar farms. Thus the family which does not have substantial outside assets cannot pay the estate taxes. So the property has to be sold and cannot be passed on to the next generation."

In spite of all the tax difficulties and other problems, inheritance possibly remains the preferred way of obtaining an outfit. I neglected to say, it's far better to inherit all to yourself rather than to be mixed up with sundry relatives who live way off someplace and who possibly would accuse you of using every known gimmick to cheat them even though you work from dawn to dusk while they watch Gunsmoke, read the funnies or march in protest to something or other.

Certain people prefer marrying a place over inheriting one. In case you belong to this school the best prospects are (1) an only daughter of a wealthy land owner (preferably a deceased land owner). (2) an attractive widow with no children or, (3) a rich divorcee. In (2) be sure the deceased husband wasn't poisoned and in (3) it might be well to determine that the former husband or husbands are well taken care of. Also, in case of (2) and (3) it is desirable that the lady be above ordinary looks. For women who wish to marry a place, I have observed that it is best to select a very old land owner, if the place is all you are interested in.

Of the possibilities in this category, I think wedding a lovely only daughter is best. Some of the great rancher-ranch wife combinations have come about this way. If she has the looks, common sense and money and you have brawn with a fair amount of intelligence, why, you're on your way. Perhaps I'd better warn you not to mistreat her, else you're apt to wind up out on your fanny.

Homesteading, although fairly popular, was an extremely rigorous undertaking. By chance you may not have kept up with this method, maybe I'd better review how it worked in the good old days. Let's say you came over from Europe on the Mayflower, Queen Mary or a raft around 1650 with little or no cash and maybe a young somewhat homely wife. Perhaps you came either because the law was hot on your tail or else you didn't like the local preacher. In those days the laws and customs allowed you to become owner of a parcel of land provided you cleared the brush off and lived there for a reasonable time. Sometimes you had to be careful not to be scalped by a hostile Indian who resented your clearing off the place on which he'd been accustomed to hunting rabbits. Fiction has greatly exaggerated the hostility of most Indians, which was no comfort to a pioneer who carelessly allowed his hair to be cut too close by an irate brave.

Later on when more and more immigrants came over and more children were born it became crowded back east with most of the land being already claimed in one way or another. So the brave people pushed farther west and homesteaded land in the middle west and far west. One of the true and interesting stories concerns the opening for homesteading of certain lands in Oklahoma. On a special date, somebody in Washington declared that anyone who wished could line up with everybody else who wished, along the Kansas border, then upon a starting signal all could make a "run" for this land. The one who got there first and staked his claim would own the land, provided that thereafter he lived up to certain regulations which according to historians were not too severe. A few who were not too scrupulous sneaked ahead during the night and were camped near the available land ahead of those who followed the rules by lining up at the prescribed starting line. These were known as "sooners", the name meaning they got there "sooner" than anyone else. The name "sooner" became thereafter applied to most Oklahomans until the "Oakie" term became common after the "dust bowl days" caused by drouth in the 1930's. Oklahoma University football teams are known as the "Sooners" so long as they are winning. When not winning the names used would not be allowed in print.

Perhaps we'd better move on so, to summarize, if I were you and expected to obtain an outfit by this method I wouldn't bother. As with grants, homesteading ceased long ago.

Stealing–Like homesteading, this method was said to be possible at one time, although as far as I know there have always been laws against it, except when stealing from the Indians. One thing I have noticed–certain governments are still at stealing land. They keep trying and therein lies the real cause of most wars.

I have never actually known of a farm or ranch being stolen in broad daylight and I suspect most yarns about it happening are pure fiction. Practically every western movie will show some villain with a gold toothpick trying to steal a ranch from a poor widder but the hero always beats hell out of the scoundrel and chases him back to Jersey City.

There are a few shysters still around who cheat on land deals. In the old days these varmits were caught and hanged by the neck until completely dead. Today they're "apprehended" and tried in several courts until finally up the line some court decides the poor fellow has been arrested without his "rights" being properly explained to him. Lawyers on both sides get fat fooling with the case. The thing gets batted around in the news media until something more shocking

such as a hippie "love-in" comes up. Finally the whole thing is forgotten and the rascal remains free.

And now we come to what is the most common way of acquiring a place—buying it. This is also the easiest way, provided you can scare up the dough someplace, or sufficient credit. There was a time not too long ago when most anybody with a few dollars, an honest face, and several note cosigners could buy land. Year by year this becomes more difficult. Even an honest clean face won't get you too far anymore—you've got to have "collateral", character references, and a batch of friends.

In case you have to start from scratch or almost that low, you might as well figure on spending the balance of your life paying for the outfit and doing without certain of the so-called luxuries your city cousins can afford. You, your wife and your kids will just have to be content with the greatest luxury of all—the great out-of-doors, with clean fresh air, beautiful sunsets; and all the other glories the Master has seen fit to supply absolutely free.

I should add that in case you have in mind leaving the place to your kids, you'd best arrange to do so while you're living else Uncle Sam will step in to take a big share when you die.

I have consulted several economists, a gypsy palm reader, our local preacher and several assistant secretaries of agriculture regarding the amount you can afford to agree to pay for land. In each case the answer has been, "It all depends". One widely read economist did go much further, he said "Well" and after several minutes pause added, "I'd sure not want to pay more'n I had to."

Summarizing all these opinions it adds up to about these.

Will the land graze one cow to one acre, or will it take forty acres? Just about anybody can figure that one out—unless the forty is over a huge oil pool or an uranium mine. In this case, I'd advise you to check into the mineral rights.

Would anybody buy the outfit in case you starve out, have to move back to town and take up your old job?

Is there an old delapidated falling down barn which you can repair for only twice the cost of a new one? Or, must you build a new one that won't have the "character" of the termite eaten one?

Apply these same questions to the house, if there is one.

Can you really enjoy nursing and tenderly caring for the land, then swell with pride when a new calf is born or a field turns rich green where once only scrub brush grew?

Can you and your good wife enjoy an early morning chorus provided by just the right mixture of voices from a robin, a mocking bird, a cardinal, a bobwhite and a host of other birds you can't identify?

Finally, do you think you can endure the greatest life on earth—out there in the beautiful fresh, clean outdoors, the smell of green growing things where you feel closer to God than at any time in your life? If you can, sign the contract. If not, forget the whole thing.

FARMING IS NICE—IF YOU CAN AFFORD IT

A good many city people think farmers and ranchers are in the process of getting rich because of high food prices and government support payments. I'm saying this without any intention of criticizing the town folks because without them the farmers and ranchers would be in one mell-of-a-hess. I suppose everyone has read about there becoming less and less farms. Farms continue to become larger and larger and fewer and fewer in number.

It seems to me there are about four types of farmers or ranchers left.

(1) The one who genuinely loves the land and who wouldn't swap that life for any other. The farm might have been passed down from his father or he might have started from scratch. In the latter case he scrimped, saved, did without, and with shrewd judgment and luck managed to pay for the place even though he was middle-aged or older when he finally did so. He makes a pretty fair living but his return on investment rarely runs over three percent, more often something like one percent—even less.

You don't believe that? Figure it out for yourself.

Let's say he owns, or is buying 640 acres of land which could easily sell for $400 an acre. That's an investment of $256,000.00 in land alone. This means he must *NET* $7,600 annually to make three percent on just the land to say nothing of his other outlays. To farm that much land he must have machinery which costs at least $50,000.00—often more. There will be two tractors, a combination corn-picker and small grain combine, a pick-up hay baler, maybe a field silage chopper, a manure spreader, plus plows, disks, planters, hay rakes, cultivator, sprayers, harrows, welding outfit, and so on. Depreciation alone on this machinery and equipment will most likely amount to over $5,000 annually. A new fence will cost at least $1,000 per mile. Then there are always repairs to make on existing fences, ponds, paddocks, roofs, barns, sheds, besides new buildings that need building. His taxes will vary depending on where he's located. A fair average might be my own farm where the real estate tax alone is something over $3.00 per acre—many places they're much higher.

One man can hardly tend 640 acres by himself so he must hire extra labor unless there are enough children, old enough, to help with the farm work and chores. Even then it costs considerable to raise children, a fact of course not peculiar to farm families. There was a time when good farm help could be had at $2.00 per day. Not anymore. Now it usually costs that or more per hour.

Investment in livestock can be almost any amount. If a dairyman, he will probably be milking at least 70 cows, an investment of perhaps $20,000 or more in cows alone, to say nothing of replacement stock, milking equipment, cooler and a grade A milking barn or parlor. If a beef man, his investment per cow will be less, but probably somewhere near the same in total dollars. Investment in other types of farming—fruits, vegetables, cash crops, poultry, hogs, will some times be less, sometimes more, but likely will average out about the same.

The family will need a car for transportation to school, church, clubs, and often more than one is needed if there are children of high school age. This means an investment of $3,000 to $6,000 for transportation plus a pick-up truck and a larger truck for hauling livestock, fertilizer, feed, and equipment.

In many farming areas the annual fertilizer, insecticide and herbicide bills run $5,000 or more annually depending, of course, on the type of farming.

As for the rancher or cattle farmer, Bill House, Kansas rancher and former President of the American Hereford Association and the American Cattlemen's Association says: "It takes at least $1,000 invested in land for each cow and you'll have to add $250 for the cow and machinery. It doesn't make any difference whether you're using land worth $700 an acre, which will handle a cow to one and one-half acres, or $20 land that requires 50 acres to the cow. In the end, the total investment in the land, machinery and the cow will run around $1,250. From this investment on the average you'll market a 400-pound calf each year."

If Bill is right, and most ranchers will say he's conservative, interest cost on the investment alone at say seven percent amounts to $87.50 to say nothing about feed, labor, veterinary bills, bull service or taxes. Bull service using one good enough to sire choice calves will amount to at least $10.00 per cow. If the cow eats a ton and a half of hay, or its equivalent, each year at $20.00 per ton, add another $30.00. Taxes, labor, vet bills and incidentals will add another $10.00. You now have at least $127.50 expense when her calf is born. If he sells at 32c per pound you've barely recovered your fixed cost, even if you had no expense in raising the calf from birth to weaning. But there are expenses! A calf will die here and there, there will be some veterinary bills, extra labor for branding, castrating, plus the regular farm expenses that go on regardless. Also, in addition to the expenses, the farmer faces uncertain weather conditions and uncertain market prices.

Perhaps this may sound like "Why battle these odds simply to make a living?" There is usually one of two reasons for farming. In one case the farmer has lived on the place all his life, loves it and would be miserable doing anything else as he is more or less removed from smog, violence, heavy traffic, muggings, hippies and all the other irritations his poor city cousin must put up with. Then there is the farmer who might sell out and move to town except he wants to leave something to his wife and/or children so if he sells now the tax bill would be murder, although he realizes that if he dies the inheritance tax could be worse. Besides, what in the world could he do if he sold out? So, he hangs on, enjoying every day of it. According to the state university economist, he is slowly going broke and would be far better off if he took a job in some factory. It's pretty hard to argue with figures. Still there's many a farm or ranch that's been paid for, a large family educated and substantial savings accumulated while the owner "slowly went broke." This group farms because they really like to do it, problems and all.

The second group is composed of those, all of whom inherited the place. There are two kinds in this group. One would like to give up the place and move to town. He doesn't because he may be one of several heirs and since the others won't sell he's obliged to stick it out, or he is not trained for any other job, so farming is the only way he can figure out to make a living, and finally the tax bite from the proceeds of a sale would mean a severe financial loss unless he invests in some other business about which he knows absolutely nothing. This second group has about the same problems as group one, except they don't enjoy farming.

There is a third group of farmers—the "investor" type. In this group there appears to be two kinds, (a) the guy who doesn't know beans about farming and usually doesn't care to learn. He's simply looking for a safe investment and more return on his money than can be realized elsewhere. This person isn't all bad by any means, as some have tried to picture him or her. Quite often he becomes genuinely interested in farming or ranching, learns a lot about it and makes commendable contributions. (b) The other kind is the "tax shelter" variety. This person is usually in the high income bracket and is attracted to farming or cattle ranching by the capital gains, depreciation and land improvement provisions of the tax laws which he's been misled into believing are different and especially favorable to cattlemen. Actually, the tax laws for farmers and ranchers are essentially the same as for any other business. I suppose the "romance" of cattle ranching has attracted more of these people than has any other business. Many wouldn't know a cow from a billy goat, but they get a great thrill from telling their friends at the club about "my cattle" on that little old Montana spread. Fortunately, there are not many in the tax shelter class for tax purposes only, but their damage has been blown up by certain politicians and journalists far above their importance. More and more city businessmen seem to be investing in farm and ranch properties This is wonderful, provided their aims are sincere and their methods untainted. Some consider it the best hedge against inflation as land becomes more scarce; some have a nostalgic memory of life in the country when they were kids, and hope to retire on the land; then considering high food prices, a few think that farming is a tremendously profitable business; and still others imagine they can save tax dollars in agriculture. The first two kinds of city businessmen are commendable and are encouraged by agricultural organizations. The businessman seldom makes a lot of money farming, and nine times out of ten, the investor for tax purposes loses more from mistakes and unnecessary expenditures on the land than his tax savings produce.

Like it or not, the purebred livestock industry has usually had and needed a certain amount of "outside" capital and as far as I can determine this is a fact not peculiar to America, nor is it peculiar to the livestock industry only. Throughout the history of American agriculture, it has been virtually impossible for a "dirt" farmer to accumulate enough surplus dollars to do the necessary experimentations, the trial and errors so necessary to make the improvements in the breeds over the years. Genuine livestock improvers have in a great many instances been men who were raised in the country, or who had some other agricultural interest, and who could afford to take the financial risks involved in experimentations.

A good many urban housewives who know little or nothing about cattle ranching and farming costs believe that the farmers and ranchers are getting rich. An animal grazing peacefully on the grass seems to be costing the rancher nothing at all to raise. The cattlemen all appreciate these housewives, for without them there would be no market for beef. At the same time, perhaps the housewives should fuss some about the cost of other household items instead of just singling out meat. How about TVs, cigarettes, toothpaste, detergents, hair nets, paper products?

Most cattlemen don't want it all, they simply want enough to make a decent living. It ain't easy!

CHAPTER 4

THROUGH GATES, OVER AND UNDER FENCES

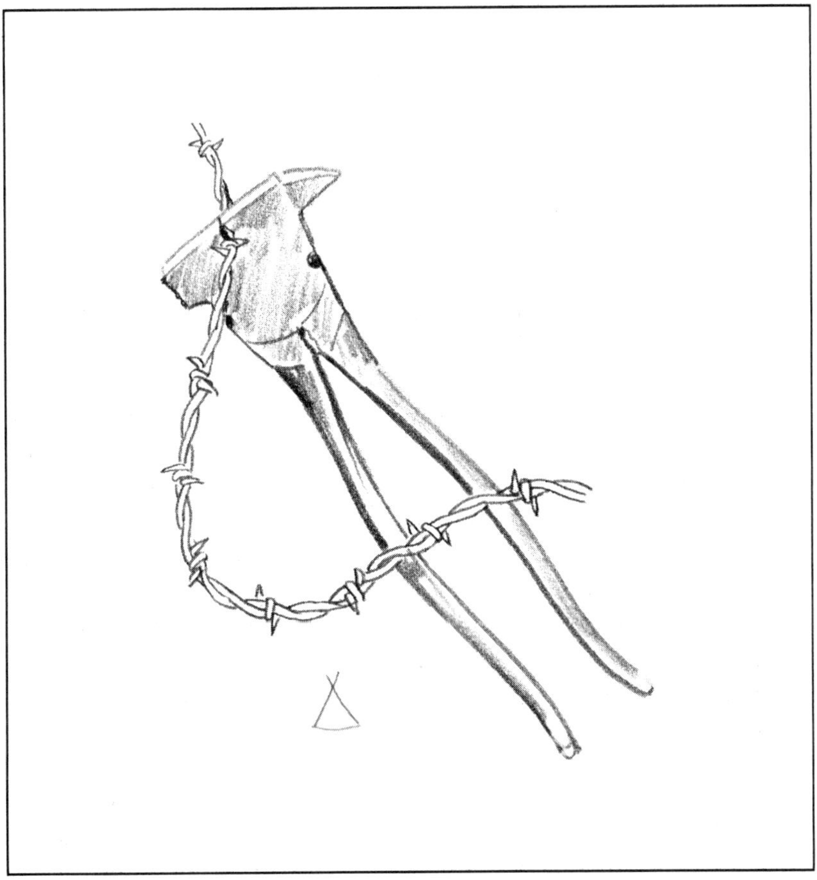

Chapter 4

Through Gates, Over and Under Fences

I claim to be the champion gate opener of the country. I have opened gates in practically every state in the Union, for all kinds of people in all kinds of weather. At present I am considered a methodical gate opener, but I'm very thorough. One time I opened a gate on the LBJ Ranch.

I started opening gates very early. The first experience I can remember was opening a gate for my Uncle Lee when I was four years old. Uncle Lee was Mamma's youngest brother. He chewed tobacco, wore size 11 shoes, and he could cuss a mule with fluency. We moved from Standing Rock, Oklahoma to a place called Boggy Depot when I was four years old, and Papa let Uncle Lee drive the mules, Rhodie and Beck, hitched to a wagon to haul what household goods we had—which wasn't much. I slept most of the way on top of the kitchen stove under Mamma's winter coat while Uncle Lee drove and sang cowboy songs, but not very well. The song came out through his nose. I remember waking up with him poking me in the ribs and yelling, "Hey, kid! Git down and open the gate." I have been opening gates ever since. Soon after that a horse fell on Uncle Lee and busted him all up. I never opened any more gates for Uncle Lee, but Papa took up where he left off.

We had a little old poor farm where Papa raised a few cows, corn, peanuts, turkeys and some sorghum cane. Mamma raised seven kids. I was and still am the oldest boy among the seven kids, and I always had to open the gates. To my dying day, I'll never know why we had so many gates on that little old farm. There was a gate to get into the place, and a gate to get out the back way, a pasture gate, cow-lot gate, mule-lot gate, pig-pen gate, dog-pen gate, garden gate, chicken gate, cornfield gate, and yard gate. There was a gate between the house and the house-out-back. It was a nuisance on a cold morning.

Since my start at the tender age of four, I have had nearly sixty years experience, I have made a study of gates and gate-openers, of hinges, gate types, gate latches, gate materials, gate weights, and the art of opening the gate with a minimum effort. I don't want to appear boastful, but I think I am qualified to classify and discuss gates with the best in the business.

First, I would like to classify gates.

Class I – Wire gates or gaps–sometimes called International gates. Practically everybody has had or has seen one of these. They are made by taking two strong hickory poles, usually about four inches in diameter and very green and heavy. As the poles dry out they should be replaced by new green ones. There are three important features of this type: (1) Poles must always be too heavy, (2) wires must be too tight, (3) latches must almost always be impossible to get off, and (4) rusty barbed wire is most often used.

Advantages of Class 1 are:

(1) They provide good workouts for a boy and teach him not to be lazy. (2) No medling woman will ever come through one–she'll try to crawl over, tear something and go on back home. (3) A smart aleck horse can hardly ever open one. (4) These gates will not obstruct any breeze. (5) You get plenty of sunlight through one. (6) You can ask anyone you don't like to open a few–sooner or later it'll kill him.

Disadvantages of this type are:

(1) There is considerable expense in continuously replacing the heavy poles. (2) An old sow and pigs can run under one. (3) A bull buyer won't come but once if you ask him to open one. (4) Doctor bills resulting from injuries in opening these gates are high. (5) No insurance company will risk a person who habitually tries to open them.

In light of the above, you can decide for yourself. I have discontinued my study of this type.

Class II – The Sagging Susie. The same general principles are involved in building this type as in Class I. However, there are some very essential differences which should be considered before attempting to make a satisfactory Sagging Susie. These are: (1) Poles should be either too little, severely bowed, or too short and broken. (Poles should never be replaced—always wrapped with some hay-wire, or put together with 20 penny nails.) (2) Wires should always be loose or better still, some should be broken. (3) Latches should be made of rope or No. 9 wire; however, latches are not too important as most Sagging Susie owners do not seem to care whether they are open or shut.

TOM PHILLIPS

The advantages of Class II are:

(1) Anybody can step over one—so can a cow. (2) There is not much up-keep, as they are not supposed to be kept up. (3) They usually last forever.

Disadvantages are:

(1) The cows will get over in the neighbor's cornfield when they shouldn't. (You don't need to worry too much about the neighbors, but it scours the cows.) (2) They don't furnish much work training for lazy kids. (3) Some people laugh at them.

Class III – Board Gates. These are usually made out of boards, although some people will argue all day that they're made out of lead. I am not quite sure how board gates ought to be classified on account of there are several ways you can do it. I knew a fellow once, his name was Buchanan, I think, who said anybody was crazy in the head to build a board gate. He said they would either warp, dry rot around the hinges and fall down, get busted up by bulls fighting through them, or termites would eat the hell out of them. I wouldn't entirely agree with Buchanan, but I do know I never saw one but what got to dragging on the front end sooner or later and nine out of ten have the hinges off in no time. There may be some way to cure this, but the only way I ever saw it done was to use a hay wire as a hinge substitute. This will work, but it looks terrible.

In spite of all their short comings, board gates are widely used and after thinking it over suppose I don't try to classify them except to say that they fall into about the following groups: new, old, heavy, heavier, bowed-in, bowed-out, painted, unpainted, oak and non-oak.

Just in case you are considering a gate, maybe I had better briefly outline the advantages and disadvantages of board gates.

Advantages:

(1) You can use nails in boards. I don't know of anything else offhand that is adapted to the use of nails. (2) You can usually buy boards most anyplace. They get a little scarce during the war years, but we aren't supposed to have any more wars. (3) By putting up board gates you have a use for all the old broken, rusty hinges laying around. (4) After board gates fall down, you can use the salvage for kindling wood. (5) Most anybody can identify one on sight. (6) They are nice for cribbing horses.

Disadvantages:

(1) Wasps build nests on them and sometimes sting innocent people. One time at home I remember the preacher came, and he had this experience—on Saturday, too. My oldest sister and I got dressed down for laughing about it in church next morning. (2) After a deep snow, you can't get one open hardly at all without shoveling a lot and the shovel is usually lost someplace. (3) A fool mule will stand and kick one for hours just for the fun of it. (4) When they are freshly painted, the paint will rub off on things—but still this might be an advantage because they look nice. So much for board gates. I am continuing my study of this type and I may have some more to report later on.

Class IV – Metal Gates. Outside of the fact that the metal gates cost more, they are fine. You can use several types of metal, but whatever you decide on, maybe you better let someone who is equipped make your gate because you might get it crooked, and then it wouldn't be worth a tinker's damn. Of all the metal gates I know anything about, the aluminum ones seem to me about the best. There are several outfits down in Texas which make thousands of them and sell them all over. Some people say they're the best thing out. I won't attempt a detailed discussion of metal gates, only to say they have the same general hinge shortcomings as have board gates. You can't drive a nail in a metal gate, they get awfully cold in winter, but they will outlast a board gate nine times out of ten.

Class V – Fancy Gates. Then there are these fancy gates where you are supposed to be able to open them without getting out of the car. These are more adapted to places where there are no kids around to get out and open the gates, for people who have arthritis in the legs and for rich people's estates. If these gates would work they would be nice on rainy days, or when you're by yourself and have on clean clothes and a shoe shine. The only trouble is, they never seem to work. They are made by putting up an elaborate Rube Goldberg contraption with an arm stuck out on each side way up in the air, and a rope hanging down on the outside end of each arm. On the lower end of the rope is fastened about one-third of an old hoe handle. You're supposed to drive up, stop the car, reach out the window, grab this hoe handle and give it a stiff yank. Through the law of gravity, a series of counter balances, and a long mathematical equation, the gate should fly open. In theory, you drive through the gap, yank on the other hoe handle and the above process works in reverse, thereby causing the gate to fly shut again. They never work—why, I don't know, but then I don't know why they are supposed to work in the first place. More research is needed on these gates, and I would suggest someone trying to interest the Atomic Energy Commission. I forgot to say that these fancy gates can be made of metal or wood or a combination of the two. If I were you and considering this type—I wouldn't.

Class VI – Electric Gates (so-called). As far as I am able to determine, these are the latest. I haven't studied too much about them because I am afraid of electricity. Every time I start monkeying around with a stray kilowatt I get bit and it's unhandy to carry rubber gloves in hot weather. You order these. Here are some of the basic requirements and characteristics of electric gates: (1) You almost have to have electricity. (2) If a cow, or a pig, or buck touches one it'll get a stiff jolt, heist its tail and go skedaddling off to the bushes thinking it's been hit by a tsetse fly or something. About two doses and they are done with electric gates forever. (3) Stray kids won't fool with them. (4) You open one by bumping into it with your tractor—or by having some city slicker get what's coming to him.

Class VII – Cattle-Guards. I have no notion why they should be called cattle-guards instead of horse-guards, sheep-guards, pig-guards or billy-goat guards. I can only say—and I am widely quoted in this statement—"Cattle-guards are sure nice." You make them by leaving a hole in the fence, much as if you were going to put a gate there. The length of the hole will depend some on whether your wife drives the car—in case she does you make it real long. Then you dig a square hole right there in the gap. I don't know how deep this hole ought to be, but be sure to make it deep enough. You make the hole, say six feet across, depending some on whether you have mules or goats—if you do, I'd suggest around twenty feet. Then you take some heavy logs and lay them crosswise of the hole. Leave them stick over the edges quite aways so's you won't fall in with a load of hay. Then lengthwise and on top of the logs you put 2 x 6s on edge, or poles, or old railroad irons, or pipe about say six or eight inches apart. You nail, bolt, or haywire these down good and you have the slickest gate you ever saw—without having a gate. Some people get real fancy and box the hole with concrete which is all right, but it costs quite a bit more—and this is pretty important in this day of high taxes.

Advantages of a cattle-guard are:

(1) Kids love them—so do older people.

Disadvantages of the cattle-guard:

(1) They are useless on a chicken farm.

I am continuing my research on gates and various other important farm problems. I will give you a progress report from time to time.

During my major study of gates I have minored in fences. However, my thesis will be confined to gates as I have several more semesters of fence research which may take a number of years considering that some travel will be involved and much filing, categorizing, indexing, altering, culling and some lying. Lying, it appears, has become quite common in some theses such as in a situation when a grade and a publication of a thesis becomes more important than facts.

According to some books and most cowboy movies, there were hardly any fences to speak of way back when pioneers and others moved west after reading in the brochures that land was plentiful and cheap. When they arrived they found Indians (good and bad), coyotes, gila monsters, saloons, wild women, poison ivy, race horse bookies and so many other things it would waste too much time to mention them all.

As near as I can determine, cow owners and sheep owners arrived out west at approximately the same time and immediately started feuding, fussing, boozing and shooting at each other from behind rocks—sometimes with perfect aim.

Some tenderfeet who don't watch TV or cowboy movies will probably not know why cowmen and sheepmen didn't like each other and each would shoot the other on sight. Let me explain it.

Cowmen claimed that sheep smelled something terrible and when one got anywhere near range grass the most awful odor was left with any decent cow refusing to touch the stuff. Sheepmen claimed cows also smelled. You wouldn't think people would fight and shoot each other over smells, but that's what they say happened.

There was another reason for this "war" which I'd probably better explain. Most sheep are smaller than most cattle. There are a few exceptions to this general rule, but if I were you I wouldn't let it bother me. Along with this smaller size goes a smaller mouth, smaller teeth, and most times, smaller bellies. Just in case you didn't study zoology in the tenth grade, a sheep is what's known as a ruminant, which means amongst other things that he has four stomachs. The same applies to cows but not humans, although judging from some human profiles you'd swear the owners had more than four.

A guy I know told me about an exam in a veterinary college where the professor asked a sophomore to name the four stomachs of a cow. He replied, "rumen, reticulum, omasum and anus." I understand the lad who gave this answer is now working in sanitary service at a small town near Muleshoe, Texas.

As I was saying, a sheep, because he is smaller and all, nips grass closer to the ground so when a gang of these little gluttons goes through a place there's not enough left for a grown jack rabbit much less a cow. So you can see this would really rile up a cowboy causing him to drink mean whiskey and load his six-shooter.

The books say it takes seven sheep to equal one cow. Upon reading this, a cowman will snort and say something like, "Hell, I wouldn't trade one cow fer a train load of the smelly pole cats." While I'm no authority on the subject, I've always felt that the 7 to 1 ratio is slightly slanted. When I was a kid at home on the farm we had a little passel of sheep, some cows, and Mamma had some chickens and sick turkeys. We also had some stubborn mules which is another subject I may get around to discussing later on. Anyhow, these few head of sheep were more trouble than all the rest. We had to clip wool off every spring with a hand turned shearer and it always fell my lot to turn the crank. Come late spring or summer, these sheep were forever getting screw worms in the south end. In spite of it all, Papa liked sheep.

Well, this cow-sheep war was one of the main reasons fences were introduced to the range country. There were other reasons like when a dude from say Boston bought up a "spread" and fenced off the water hole at which everyone had been accustomed to watering his stuff—cowmen and sheepmen alike. Along about this time wire cutters were invented.

The early fences were meant to keep the cows in and the sheep out, I suppose. They accomplished the former but not the latter. The sheep simply crawled under. Do you understand this? Early fences didn't especially reckon with sheep "poachers." A poacher as explained to me by a cowman was a no-good son-of-a-bitch who owned a band of sheep, drove them up and down the country, mostly grazing on right-of-ways and if they crawled under the fence and on to your outfit, why that was your tough luck.

Later on somebody invented woven wire and the range war was over except for a minor skirmish here and there. Today, cowmen and sheepmen get along like double first cousins. Sometimes a ranch owner will have both cows and sheep, but he hardly ever brags about it.

I've crawled over, under or through about every kind of fence that's ever been built, in about every state in this country and a few foreign countries. I've torn pants, shirts, gloves, hands and drawers. I'm a mess of large and small scars from end to end. So I thought you'd be interested in my observations on fences up-to-date, although I've already admitted that my studies are not yet completed. I will herewith name the classes of fences: rock, rail, board, barbed wire, woven wire, combinations of barbed wire and woven wire and electric.

Rock Fences. Practically everyone has seen a rock fence somewhere sometime. However, since they may be a few of my readers who have not, perhaps I'd best devote a reasonable amount of time to them. Rock fences are normally found in rocky country. Hardly anyone bothers to build them anymore. There are several reasons why: (1) Kids and older people no longer get a thrill from picking up rocks and transporting them to a suitable place to build a fence. (2) Hardly anyone around anymore knows how to stack rocks so they won't fall down. You can overcome this by using cement mortar, however, this increases the cost somewhat and the way taxes are nowadays, who can afford mortar? (3) There are some other reasons but I can't remember off-hand what they are. The above two are enough anyhow to rule out most rock fences. Every now and then, someone puts a fancy rock wall around an estate, a graveyard, or park, but rock fence building is becoming a lost art.

I have never seen a book, thesis or other written material devoted exclusively to rock fences. Maybe someone ought to do this.

Board Fences: Board fences are popular in the race horse country and in certain other places. I'm not sure, but I imagine board fences became fashionable before wire fences, but after rock fences. If you're curious about that sort of thing, you might go to the library and look up the respective dates. I couldn't care less. Some of the advantages of board fences are: (1) They look real nice when lately and appropriately painted. White is the most common color, although you'll see green, yellow, red, chartreuse, and several colors difficult to identify. (2) There's absolutely no chance of a young scampering colt getting wire cuts on a board fence. True they may try to jump over, fall down and break a leg or neck, but no wire cuts. (3) Board fences are much easier to crawl over than are wire fences, yet they're usually much harder to crawl under. Since you might be thinking about building a board fence you ought to be alerted to some of their disadvantages. (1) Unless they are kept painted they look like the very dickens, in which case you're constantly having to apologize for them with such excuses as "Really, we think they look more typically rural unpainted," or "The fellow I had hired for the job got drunk, ran off with this widow and we ain't seen hide or hair of him since," or "The kids were supposed to do the job but they got tied up with school and dancing lessons," and so on. (2) Termites get fat on a board fence. All manner of stuff has been pawned off on board fence owners designed to kill, discourage or drive away termites. Perhaps some may work under ideal conditions—however, no conditions seem to be ideal. (3) Nearly every board fence you run into has a few boards scattered around that have fallen off, been kicked off, butted off or rotted off. With board fence carpenters hard to come by anymore, this can get to be an unsightly problem. (4) Hardly any busy ordinary short-in-cash dirt farmer can afford a board fence.

Rail Fences: According to some historical writers, Abe Lincoln did more to popularize the rail fence than most anybody you can name. I have always felt that there is quite a lot of fiction in the stories about Abe splitting rails all day long, then studying reading, arithmetic, and politics by candle light far into the night. Be that as it actually was, rail fences are considered charming by a certain class of present day architects, estate owners and women tourists. Through the Smokey Mountains in places like Virginia and North Carolina, you will still see quite a few, but there seems to be no special effort to replace them once they fall down. Just in case you've never seen a real rail fence, the old-time ones are zig-zagged. Some more modern ones are laid straight.

I doubt if rails ever replace wire for fencing anymore, unless some way of building them by computer is thought up.

I forgot to say that rails are usually of oak timber, but I suppose most any kind can be used except hedge—anyone who has tried to split a hedge log will understand this. Out West there are quite a few corral fences made of rails from the local woods such as pine, cottonwood or aspen. A few hillbilly songs have been written about "Leaning on the Old Rail Fence" and some are real pretty when sung by the Sons of the Pioneers or Eddie Arnold.

Electric Fences: As the name implies, electric fences require some system of providing an electric jolt. Electricity has been known since Ben Franklin's time, but its use in fencing is of relatively recent origin. I intend to look up the exact time right away. Most any farmer knows all there is to know about electric fences, but in case you're not a farmer, maybe you'd like to know something about them—you might want to buy a farm some day in which case I'll advise you to know a real liberal banker. When you find one let me know.

Most animals are exactly like humans in their distaste of an electric shock, and therein lies the basis of an electric fence. For the uninformed, you string one wire around whatever you want an animal to stay out of—or in. Every now and then hitch this to a post, pole or rod stuck in the ground. You have to have insulated clips to attach the wire to these posts. This is in order that the electricity doesn't go off into the ground directly instead of through the cow first when she touches the wire. Now, after you've got the fence up, you hook a hot wire to it. The "hot" can come from a battery or from a wire plugged into your regular line. It doesn't make much difference what the source is, just so it's fairly strong. Batteries use up their juice and, of course, must be replaced now and then. The electricity doesn't run through the wire constantly, but it goes in spurts so arranged by a "breaker"—whatever that is.

Most animals respect electric fences, keeping away at a reasonable distance once they've felt the sting. Occasionally you'll run into a beast that has absolutely no respect for an electric fence. They're either so insulated that they get no charge or else they just don't give a damn—I don't know which.

I've seen thousands of electric fences—maybe it's hundreds—and I've had about three years personal experience in using them. This will no doubt make some folks fighting mad, but I simply don't care much for electric fences. I expect I just didn't get mine put up right or something because most people swear by them. Here was my trouble. I couldn't seem to get mine high enough that an onery cow wouldn't hop over, or low enough that a runt calf couldn't crawl under. Then another problem was when the weeds grew up under them they were constantly "shorting out." Oh, I know the book says to keep the weeds cut from under an electric fence, but I was always too busy at some unimportant thing.

I meant to tell you about my wife grabbing this electric fence—she later claimed she didn't know I'd turned on the juice. I never quite believed that, for what did she think it was doing up there in the first place? Anyhow, she let out a scream you could hear for approximately a mile and a quarter and accused me of trying to electrocute her. She didn't settle down until about four days before her next birthday.

I am discontinuing my study of electric fences—also rock fences and rail fences. Henceforth I will direct all my efforts on board fences and wire fences. I will spend only a small amount of time on board fences.

Wire fences are far and away the most commonly used and I imagine they are not apt to be replaced any time soon by any other kind. I have not looked up the date when wire was first made, but I'm sure it was sometime ago. There are some hobby wire collectors around who have all manner of weird barbed wires. If you get time, you ought to look up one of these fellows and view his collection.

I will make about three general categories, with appropriate subclasses of wire fences.

1. Woven Wire
    - a. for cows only
    - b. for pigs only
    - c. combination cow and pig
    - d. for chickens and rabbits
    - e. for rats and other small pests
2. Barbed Wire
    - a. 1 wire (not recommended)
    - b. 2 wire (seldom recommended)
    - c. 3 wire (only some-times recommended)
    - d. 4 wire (ok if you can't afford more)
    - e. 5 wire (satisfactory for most—excluding rodeo stock)
    - f. 6 wire and up (expensive but usually satisfactory and very nice to look at)
3. Combination Woven and Barbed Wire

    This is made by having woven wire on the bottom and barbed wire on top. The woven wire stops small animals such as sheep, pigs, goats and mean kids, while the top barbed wire discourages cows, bulls, steers, stags, heifers, most horses and meddling women.

You go ahead and put up whatever kind you wish and can afford. On my own farm I've put up four- and five-wire barbed wire fences. There are several reasons why I've done this.

1. I can't afford any more.
2. You can forget the other reasons.

After paying for several miles of fencing, I'm coming out in favor of "open range," but I doubt if I'll get far with as many wild automobile drivers as are around these days plus the soil bank and that sort of thing.

By now you can see that this discussion of fences is strictly for the uninformed. In another year or so on my farm I'll likely do a piece on "Repairing Broken Down Fences."

©

CHAPTER 5

CROSSBREEDING—O.K. IF YOU QUIT SOON ENOUGH

Chapter 5

Crossbreeding—O.K. If You Quit Soon Enough

Crossbreeding in practically all species of plant and animal life has long been practiced by scientists and practical farmers and ranchers. To deny that crossbreeding has some value would be ignoring the facts. Those who are most vocal in its praise seldom mention its shortcomings.

In the cattle industry, the ones who are usually most lavish in praising crossbreeding are those who wish to sell their bulls to someone whose cows are of a different breed and more numerous than their own breed.

Most anyone familiar with research information will know that the first cross (f1) offspring will almost always respond somewhat better than will straightbreds under the same conditions. The important fact, TOO OFTEN FORGOTTEN, MISSED OR IGNORED, is that this is true ONLY when both parents are of acceptable conformation and quality—and prepotent enough to pass those characteristics on to the offspring. Otherwise the cross is apt to produce the poorest characteristics of each parent. This can happen even when both parents are considered superior.

To illustrate, Wilbur Plager, executive secretary of the Yorkshire Swine Association, tells this story about crossbreeding in dogs:

Seems this fellow had a dog who couldn't fight a lick nor was he swift enough to run when attacked by another dog. Consequently, he regularly got chewed up by any mutt that happened to come by. A neighbor on one side had a bulldog who was vicious as a fighter but slow on foot, while another neighbor had a cowardly greyhound who could run like a deer. So this fellow decided he'd get a bulldog-greyhound cross, thereby obtaining a dog that was fleet of foot but could also lick most any other dog when attacked. So, when he got the cross—the result was—a dog that could run like the bulldog and fight like the greyhound.

Most shortcomings of crossbreeding can be overcome by careful selection and excellent management. Maintaining good stock animals with proven performances is much more difficult than is a sound program always using cows and bulls of the same breed.

Hundreds of articles extolling the value of crossbreeding have been published. Thousands of lectures have been made and many experiments have been conducted regarding its value. In these, not often are the shortcomings mentioned. The average farmer is lulled into dreams of superior performance by crossbreeding.

Listen to what Dr. Rufus Cox, highly respected, now retired, head of Kansas State University's Animal Science Department, has to say on the subject:

"There is such a constant volume of promotion, both verbally and through the press, of crossbreeding that beef cattle producers are continuously requesting factual information on its advantages and disadvantages. And crossbreeding does have both. In spite of this, most of what is said and published presents only the advantages, with no mention being made of its limitations and real costs.

"Furthermore, most of the promotion implies that the practice is new—that the few who have tried it are just discovering in it a great new source of additional revenue without any additional costs being involved. It is further implied that a wave of crossbreeding is sweeping the country, and that in a matter of a very short time those who do not 'get with it' will be at such a disadvantage that they will be unable to compete with those who have 'seen the light.'

Sincere but Perhaps Misinformed

"Most of this is undoubtedly sincere and well-meaning. But, also, most of it obviously is based on a lack of complete information, or, perhaps, in some cases on misinformation.

"Just how new is the practice of crossbreeding beef cattle? What has been its history, and what of the experiences of those who have tried it? One of the earliest experiences in crossbreeding by producers of high-quality beef cattle to be widely publicized occurred somewhere around 125 years ago. Beef cattle producers in Ireland bred

 Angus cows to white shorthorn bulls to produce the famous Irish blue-gray 'store' cattle (stockers) which were in great favor by Scottish and English feeders who fattened them in stalls, largely on roots (mangolds or stock beets), oat straw and linseed meal. There was no denying their superiority over pure breeds, for the feeder.

"To jump quickly over a period of years, crossing of the beef breeds received some attention from American cattlemen soon after the English breeds became established in this country. At this point is should be recognized that the practice of grading-up by using a purebred bull on grade cows of unimproved breeding is a vastly different thing from crossing two improved breeds. Periodically over the years from these early beginnings up to the present, there have been waves or cycles of a few years of growing interest, followed by periods of receding interest in which the practice all but disappeared for a time. Just now we are in the midst of another upward surge in which the flame is again being fanned.

Basic Laws Need Be Appraised

"Temporary and permanent results from crossbreeding should be surveyed, and the basic and practical genetic laws appraised. The FT_1, or first-generation hybrids, usually are superior to either parent in several respects. This is a result of what is known as heterosis, or hybrid-vigor. Calves of this composition usually are larger, grow somewhat faster, gain more efficiently, and often (not in all instances) have more constitution. In some cases they are more hardy than one parent breed and less so than the other.

"Likewise, if the hybrids are bred, possibly more often than not they are more fertile, but here again in some cases may not be more fertile than both parent breeds. Claims for other advantages, such as better carcasses, for example, have been made, but it is questionable whether this has been substantiated by unimpeachable carcass studies and measurements.

"Too often our story ends here without any effort at presenting the other side, or even recognizing that there is one. Let's again turn briefly to practical applied genetics. Pure breeds were established, after first crossing many unimproved varieties (not breeds) in order to induce numerous variations, from which to choose. Subsequently, early-day breed-builders turned immediately to 'straight' or 'pure' breeding to reduce variations, and to 'fix' type, color, size, carcass and numerous other characteristics. This served to establish animals (breeds) which were prepotent, that is, which possessed the consistent ability to pass their traits on to their immediate offspring and succeeding generations. Inbreeding often was followed to intensify and hasten the fixing of characteristics.

"This process, which admittedly is over-simplified for the sake of brevity, produced our pure breeds which breed true. Obviously it makes no difference how good a beef bull is, or how many good traits he has, if he cannot transmit these characteristics consistently to his offspring he is good only for slaughter. This is an example of the immediate, but temporary, improvement to come from crossbreeding.

Practical Problems for the Breeder

"The breeder who decides to embark on a crossbreeding program almost invariably points to the premium he believes he can get for crossbred calves over those of pure or straight breeding, then follows with the reasoning that he can buy a bull of another breed as good as one of his own breed, and at a comparable price. He is right, up to that point.

"What he doesn't foresee are the problems to follow. On the average, replacements in the cow herd average 17 to 20 percent annually. It is almost impossible to cheat on this figure, and the tolerance below it is narrow. Failure to replace any at all the first year will mean double that number in the second year, and so on, until in five or six years all will need to be replaced. If a lower replacement level than approximate percentage indicated is followed, one simply does not have a good culling program and is not improving his breeding herd as he should.

"Even so, he is faced with the replacement problem. In general, he has two (maybe three) alternatives.

"First, he can breed part of his cows straight (within the breed) and crossbreed the others, thus producing his own replacements. Can the average producer afford all the extra lots, pastures and cross-fencing of pastures, as well as extra help, required for this? Only if he is an extensive operator with a far larger operation than average.

Simple—Until It Is Tried

"Second, he can crossbreed all his cows, and buy all replacement females. This may look simple—until he tries it. Then he finds it is very difficult to buy heifers of the quality he wants, all of similar breeding, and at a price which will leave him a chance to succeed.

He generally will have to take the second cuts, after the breeder has retained the tops, which, of course, every good breeder must do. It is rare when a man can buy 25 years of breed improvement. And, incidentally, he will already have spent far more than the extra premium for which he sold his crossbred calves.

"The third possibility is to retain the crossbred heifers for breeding. He has been told, rightly, that hybrid vigor in these heifers usually (not always) will make them better milkers, will result in a larger calf crop and possibly in another advantage or two. He has not been told that in the F_2, or second generation, hybrids, the hybrid-vigor generally not only has disappeared, but that the quality level has receded below that of the original straight-bred grandparents.

"Moreover, genetic behavior again has taught us that this process produces the exact opposite of inbreeding, or even ordinary within-breed breeding. The result thus is a wide dispersal of characteristics and the appearance of all sorts of sizes, types, color combinations, etc., with extremes in both directions as well as intermediate kinds. Furthermore, for every visible trait to show up, one can be sure that there are even more numerous ones present which cannot be detected visually, such as widely varying growth potential, gaining ability, feeding efficiency, loin and rib eye size and other carcass traits, fertility, disease-resistance, milking ability, and so on and on.

"Following this pattern of breeding for a few generations, possibly not more than four or five, will result in a herd that is a heterogeneous conglomeration of types, colors and sizes, and which carries many widely varying traits that cannot be detected visibly. It is questionable if they are any improvement over the cattle with which breed-building was started many generations ago.

"What if all breeders followed such a breeding plan! Where would we get cattle to cross breed?

Heterosis Available Within the Breed

"Furthermore, breeders can get heterosis by mating widely-related strains or bloodlines within their own breed just as effectively as by crossbreeding, and do it without inducing numerous, widely fluctuating and uncontrollable varieties of contrasting characteristics which swing from one extreme to the opposite, with all degrees of intermediate variations between.

"The third of the three possibilities mentioned earlier undoubtedly is the worst of the lot. In the long run, this kind of a program is a down-grading and not an upbuilding breeding program. We cannot violate the laws of genetics. We must accept them, work with them, and let them help us. So let's look to and harness for our own use these inviolable laws of genetics and use with them a program of practical beef herd management as we watch carefully the out-go and the come-back of the dollars on which we are operating the business.

"Now, one other angle! How serious is this problem and to what extent do we need to concern ourselves? It is doubtful that even with all the promotion, the percentage of our beef cattle which are involved in crossbreeding is any higher today than it was generations ago. For its perpetuation it depends almost altogether upon a new crop of cross-breeders all the time. There are a few exceptions to this, but for the most part those who go into such a program abandon it when they experience the replacement problem. Thus it appears to be self-limiting.

"Crossbreeding involves far too small a segment of our beef cattle to be a threat to any breed. But it does give quite a setback in time and dollars to many good breeders who hear or read only the favorable aspects before involving their herd in it.

"In summary, honesty compels one to recognize the advantages to cattle feeders and perhaps to some others to be derived from the crossbreeding of beef cattle. Furthermore, it is recognized that it has made great contributions in specialized cases such as imparting resistance to certain diseases such as Texas fever, and to extreme heat.

"But heterosis, like anything of value, does not come free. For the average American beef cattle breeder, it costs more than it is worth."

A few years ago, the National Association of Livestock Record Associations put out a little pamphlet titled "The Case for Registered Livestock." On crossbreeding the pamphlet had this to say:

"Much has been written and said about crossbreeding, which is simply the mating of individuals of two distinct and separate breeds.

"Those who maintain that there may be some 'first cross' advantage usually neglect to point out that, without registered, purebred, parent stock, crossbreeding would be impossible. There just must be, if one is to crossbreed, ample breeding stock continuously available from two or more of the pedigree-recorded breeds. The initial advantages of crossbreeding are a dividend of the constructive work of many purebred breeders over long years in establishing a concentration and uniformity of germ plasm carrying performance characteristics. Without registered purebreds there is nothing to crossbreed, and thus even to the crossbreeding proponents registered purebreds are a vital must.

"It is generally agreed that most commercial advantages of crossbreeding, if any, accrue in the first generation of crossbred progeny. This being true, the crossbreeder is faced with a continuing problem—replacement breeding stock. There is serious question whether there is any advantage in crossbreeding after the first generation, but if

there is any, it can only be achieved by 'criss-crossing' which necessitates the economic problem of maintaining three kinds—two purebred and one crossbred—or else depending on a purebreeder to furnish seed stock—a system which foregoes selection by observed performance—plus added expense. Great breed improvers have historically relied on 'within herd' observations as guides in constructive improvement. It is doubtful whether this will change.

"At best, after the first cross, experience in time will prove that even bringing in the third breed in a 'crisscross' program, the commercial breeder will be fortunate if he just holds even in retaining crossbreeding advantages. If this be true, what did he gain? Merely the dubious privilege of again doing something different. Whereas, had he stayed with a pure breed, measured the performance of his

 individual animals, and kept enough breeding (pedigree) records to permit judging the genetic sources of the best performance and the best quality, he would have been better off. All this is particularly true for the owner of a small to medium size operation. Crossbreeding at its dubious best demands extra capital and operating expenditures for herd management if there is to be any chance of success.

"To discard whole herds and start over with premium-priced pure breeding stock to start a new crossbreeding program every few years cuts into a producer's profits and increases his financial hazards. The so-called 'little man', still the backbone of our agricultural economy, just can't stand it financially. He can, however, stay in business profitably by using a pure breed. He can limit his future expenditures for breeding stock largely to herd sires when needed, and to the occasional individual or small group of purebred females."

Another one who has had many years experience in the academic profession, in the Agricultural Extension Service, in purebred cattle raising, feedlot operation and commercial ranching, is Charley Kyd of the Kyd Cattle Company at Three Forks, Montana. In a down-to-earth discussion of the crossbreeding subject published in the July 1, 1967 issue of the *American Hereford Journal*, Charley titles his article "Crossbreeding-Third Time Around." . . . :

"For the third time in my life, now, I've seen an increase in crossbreeding, and probably the most vociferous noise about it this time is from the college people.

"The first cross does give some vigor and a marketable calf, but with less uniformity in the crop. I am not sure that the vigor is much greater than can be had by crossing two lines within the breed, especially when two good lines are selected.

"Almost all feeder cattle sell to a considerable extent through eye appeal. The recent propounding of continuing crossbreeding, and never going back to straight-bred cows, doesn't worry me too much, because it won't work.

"Much of the crossbreed promotion is based on the theory that the best in both parents will pop out. Put a third one on, and some say it will still happen. Put a fourth one on, as some are now advocating, and look at the possibilities—all of those good features in one animal. There is just as much possibility, if not more, that the poorest qualities of each will pop out, all wrapped in the smallest animal.

"And consider the risks: uniformity is gone, top prices as feeders is gone, selection of uniform loads of fat cattle is impossible and the breeding herd is 'shot'. One certain result would be that the demand for Hereford females would go up.

"I have some friends in the commercial Hereford business in Mexico. They are among the strongest proponents of straight bred cattle that I know. The Mexican Coriente is a bovine that is the result of crossbreeding. They can present some interesting and revealing figures as to where profits come from, as between the two kinds.

"Even the first cross, and then only on first-calf heifers, can result in maintaining the quality of a range herd. The demand for replacement heifers is great, even in a well-managed herd under range conditions.

"A 90% calf crop is considered good in a range herd. The good operators sell all cows that don't raise a calf. So there goes 10% of the herd. A death loss from all causes in a 12 month period, plus those culled for unexpected things will easily go to 3%. Under normal range conditions cows should be culled at nine years old, and that will take another 7%. So you need to save back at least enough heifers to replace 20% of the herd each year. If you cull also for efficiency you can boost this figure a little.

"With a 90% calf crop, only 45% of the cows will have heifer calves. If you are crossbreeding all the first-calf heifers, or 20% of the herd, only 35% of the cows will have straight-bred heifer calves, and to get needed replacements 57% of the heifer calves would have to be kept. Most range outfits would then be in the position of making two cuts. First, a cut of the heifers the owner wants to keep, and second, a cut of some more he would *have* to keep. Quality would suffer."

The only conclusion about crossbreeding I have ever been able to reach is—The first cross, using acceptable parents, will produce pleasing results. After that you'd best get busy with the many, many other good management practices that guarantee even better results or, if you prefer, the scripture Leviticus 19:19 reads as follows: "Ye shall keep my statutes. Thou shalt not let thy cattle gender with a diverse kind; thou shalt not sow thy field with mingled seed: neither shall a garment mingled of linen and woolen come upon thee."

CHAPTER 6

DWARFISM—ITS SOLUTION

Chapter 6

Dwarfism—Its Solution

Nothing in recent years has given American cattlemen greater concern than these two; beef imports and dwarfism. Beef imports is primarily a political matter and its discussion will be left to those who pour over such things. Dwarfism, however, was, and is, a different matter.

Back in the 1940s and '50s dwarfism in cattle was just about the hottest issue around. Today, most cattlemen know what is meant by a "dwarf". In those days, many didn't. Authorities on the subject were a dime a dozen, some of whom wouldn't have been able to identify a dwarf if they'd met one on Main Street. If you are young enough not to have lived through the hullabaloo or in case you're an older prospective cattleman who knows absolutely nothing about the whole thing, perhaps I'd better give you a little review.

Webster's dictionary defines a dwarf as "an animal or plant much below the normal size of its species or kind." This definition is fine as far as it goes but is hardly adequate to describe a snorter-type dwarf calf.

When one is born, it fairly well resembles what a normal calf should be. In fact, even if you're accustomed to seeing dwarfs you will probably think your calf is normal. Soon, however, the critter starts to look like something has gone wrong. Its breathing begins to resemble that of a very old man with asthma following the chase of an unruly heifer. The eyes start "bugging" out and it stands around with its tongue hanging out—just a little at first—then more and more as time goes on. Instead of all parts of the body growing in the proper proportion, the belly and head grow most, with the rear end and length of leg lagging far behind. If it lives beyond a few weeks the stunted appearance becomes more pronounced. Its gait becomes wobbly and its breathing more like a snore or snort—hence the name "snorter" dwarf.

Its growth continues in the wrong places with the bulging belly, goofy head and peaked rear becoming more and more pronounced. Finally you get so sick of looking at the thing, you'll either knock it in the head, send it to the butcher, where they may use it for hamburger, or you'll donate it to the local experiment station for "tests".

According to the literature, dwarfs have been known in all animal and plant life for lord knows how long, but it didn't become a serious problem in cattle until about the 1940s. I personally recall seeing one in 1932, but not another one until 1946. Just why the sudden alarming increase should have taken place around the late '40s, no one has yet satisfactorily explained. Anyhow, the usual solution by a breeder when the first one or the first few showed up was to knock the things in the head and conclude it was just "one of those things." Various theories for its cause were advanced, among them atomic fallout, too little rain, cattle becoming too small, poor feed conditions, an act of God and a host of others. Breeders would huddle in small groups behind the barn and discuss the matter with considerable concern, but few admitted publicly that they themselves had had any.

As the incidence became greater and greater and discussions became more open, the board of directors and Jack Turner, then secretary of the American Hereford Association, decided that if possible, something ought to be done. L. P. McCann, who was at that time, fieldman for the American Hereford Association and who had had considerable experience with academic matters, public relations and digging up stuff, was assigned to the job of accumulating sufficient information to attempt pinpointing the cause. Mac had full support of the directors and Jack, but naturally not all breeders.

So Mac lit out, traveling by plane, bus, auto, hitch-hiking and sometimes on foot, buttonholing about anybody who'd stop long enough to discuss the subject. Around the AHA office he became "Runt" McCann. The problem had gotten so serious that most Hereford breeders were willing to talk. Some said they'd had many dwarfs, some a few, and some said nary a one.

On one hunt, Jack Turner, several board members and a few others made what became known as a dwarf festival (tour) stopping at many breeders' ranches or farms to discuss the subject. Most breeders freely discussed their experiences, supplying Mac with pedigree information on animals in their herds that had produced dwarf calves. Mac also wrote hundreds of letters, made a raft of expensive phone calls, and finally accumulated an enormous stack of information about "guilty" animals.

This information he toted around to leading geneticists of the time; among them were Dr. Jay Lush and Dr. Lenoy Hazel of Iowa State University, Dr. Doyle Chambers of Oklahoma State University and Dr. H. H. Stonaker of Colorado State University. The idea was to attempt to find out if dwarfism was hereditary or was there some other cause. If hereditary, how did it behave and how to go about clearing it up?

This book is not the place to report all of the time, analyzing, discussion and experiments required to come up with the conclusion. I will just say that it was determined and proved—that "snorter dwarfism" is the result of a "recessive" dwarfism gene and appears only when both the sire and dam carry this dwarf factor.

Genes behave in a crazy way sometimes and a fellow by the name Mendel determined long ago that you can't always depend on them to act just right. A simple recessive gene works fairly constant provided you consider enough cases. On the average it works like this with, say, 1,000 cases—a 1—2—1 ratio.

Normals	1 —	250
Recessive	2 —	500
Dwarfs	1 —	250

Do you understand this? I don't quite understand it either, but it works that way anyhow.

Since the hereditary simple recessive theory proved to be correct, the next job fell into place—some way to eliminate the gene for dwarfism from the breed.

There wasn't, and still isn't any way to distinguish between a normal animal with "recessive" genes for dwarfism and a normal one without these. Numerous experiments were conducted in an attempt to find a method of identity; x-raying, testing the blood, shooting with chemicals, measuring and a host of other methods. None proved worth a quarter.

Meanwhile, the information painstakingly assembled by McCann definitely pointed to certain families and certain animals within those families that were more guilty in producing dwarf calves than in other families. The information strongly indicated and later proved correct, that some families were "clean". 73

The directors of the American Hereford Association ruled then, that the information should be made available, with discretion, to those who requested it.

Even those not familiar with the purebred cattle industry can well imagine the storm this created. Good breeders, some of whom had spent a lifetime building a herd found their cattle more or less condemned because animals related to theirs had been reported as dwarf producers. A number of breeders lined up "against" the Association's policy of attempting to help breeders out of their dwarf dilemma or to help others to avoid it. A group called the "Committee for Hereford Progress" was formed and a platform was prepared, one plank of which demanded that the Association cease "analyzing" pedigrees on the basis of information supplied by accumulated data. Some of this group were sincere in their belief that the information was inadequate or inaccurate, resulting in animals and whole herds being condemned without adequate proof. Certain other breeders, both in and out of the "Hereford Progress" group, simply wished the Association to continue accumulating the information but to keep quiet about it until they had a chance to work out their own problems—on occasion by selling off questionable animals to unsuspecting newcomers.

Jack Turner retired as secretary of the Hereford Association on January 1, 1955, and I inherited the job.

1955 was a stormy year with much bitterness and considerable name-calling by breeders on both sides of the issue. The Association was "caught in the middle" with demands coming from breeders who held opposing views. One very vocal breeder threatened to sue the Association for all its assets if the so-called "pedigree analysis" was not promptly stopped. Fortunately the Association's directors stood fast in their determination to adhere to whatever steps were necessary to eliminate dwarfism from the breed. However, the board also wisely ruled that such steps had to be supported by documented information.

The 1955 annual membership meeting drew a larger attendance than had been known for many years—perhaps ever. The "Committee for Hereford Progess" had been active in urging breeders to attend with many who opposed the Committee's views likewise urging attendance. During the late summer the mails were full of pleas, propaganda and individual letters. Some were extremely bitter. The "Committee for Hereford Progress" held a few meetings in which it developed a "platform", the same being widely distributed to association members. Directors of the association likewise held a meeting in which they framed a reply to this program.

Naturally copies of each fell into hands of competing breeds and representatives of the press. Certain publications which reveled in a squabble found the Hereford plight a source of much glee. Other

TOM PHILLIPS

 breeds remained quietly apprehensive, but without doubt gained considerably in increased numbers by reason of the commercial breeders attempting to "work out" of the problem or to avoid it by switching breeds or by crossbreeding.

The American Hereford Association policies are determined by a board of directors consisting of nine breeders who serve three-year terms. By custom, a director may succeed himself for another three-year term, but must be so elected by the membership. At each annual membership meeting, therefore, three new directors are elected—or current directors re-elected on the "revolving" board. It is also a custom for the president of the association to appoint a nominating committee in advance of the annual meeting. At this 1955 meeting, the nominating committee proposed the names of three breeders, two of which were already board members. The "Committee for Hereford Progress" nominated three other breeders. Prior to the meeting word had been passed around that the "Committee" would nominate three, and many letters, telephone calls and individual visits were made urging support for the "Committee's" slate.

While the "Committee for Hereford Progress" platform included a number of "planks" the real issue was the association's program for elimination of dwarfism from the breed. Other planks were simply window dressing.

At any rate, following a rather lengthy meeting spiced with considerable name calling and not a small amount of bitterness, time came to vote on the nominees for directors. Directors proposed by the usual nominating committee were elected by a whopping 12 to 1 majority. In effect this result clearly meant a mandate to the association directors and management to strongly proceed with its efforts toward eliminating dwarfism from the breed.

The board did, however, wisely determine to conduct another breeder survey to substantiate its previously collected information and/or add thereto. This was immediately done, with only minor exceptions to the accumulated information. Certain Hereford families thought to be "clean" of the dwarfism gene continued to lack dwarf performance. Soon thereafter any animal certified as having produced a dwarf calf was so recorded in the association's computer system.

Breeders were advised that the association would upon request by the owner of a subject animal, have its pedigree automatically checked through six generations. The check would list any animal in those six generations that had been officially reported as having produced a dwarf calf. In addition to those reported through the breeder survey, several hundred previously known dwarf producers were recorded.

A considerable number of breeders voiced a strong disapproval of the ownership requirement for a dwarfism check. The AHA board reasoned that to hand out such information to other than owners would constitute a breech of the long established and correct policy that any information in the association file regarding an individual breeder's cattle belonged only to the owner of the cattle concerned. Breeders wishing to purchase herd bulls or females argued that they should be supplied such information in order to avoid "unclean" animals. The logical reply by the association management was "require the breeder from whom you propose to purchase to obtain the information from association files." The procedure usually worked well although a bit more trouble.

Although the association's procedure was responsible for almost completely eliminating dwarfism from the breed within a short few years, even twenty years later some will argue that the program was biased, dictatorial and unfair.

Perhaps this account of the dwarfism solution may read as if the American Hereford Association was solely responsible. This is far from the intent of the discussion or the facts. L. P. McCann, in an article titled "Mission Accomplished", which appeared in the *American Hereford Journal* on July 1, 1961, makes this clear by the following statement:

"In the successful pursuit of the dwarf problem, the animal research scientists of various universities and experiment stations have contributed their time and capabilities to a degree which cannot be measured. Without their help this successful campaign against dwarfism never could have been waged. Although the matter of pedigree analysis for dwarfism in herds has undoubtedly been less than 100 percent perfect, it has set the Hereford breed far in advance on the dwarf problem, thanks to a far-sighted Board of Directors 12 years ago."

CHAPTER 7

"HOLLOW-TAIL" AND "HOLLOW-HORN"

Chapter 7

"Hollow-tail" and "Hollow-horn"

None of the animal health books or the veterinary college courses seem to include any discussion of what was once considered two of the most dreaded cattle diseases, namely "Hollow-Tail" and "Hollow Horn."

You may be too sophisticated or too educated to believe that either of these two diseases ever existed. If so, you surely weren't raised around Warner, Oklahoma. Nearly all the natives around there who owned a family milk cow or any other kind believed that one of these two diseases was responsible for all sick cows. Non-natives who moved in from Arkansas or upper East Texas and who may not have believed in them soon became convinced or else they moved on, since the Warnerites were rather convincing in their arguments often to the point of challenging a non-believer to a fist fight over the issue.

My father believed that there was no such thing as a Hollow-tail or Hollow-horn, but he kept quiet about it since he was a rather small man. In addition, he was a part-time Baptist preacher and could ill afford to antagonize any of the congregation.

Another non-believer was Dutch Schaublin, who was the local veterinarian. Dutch's veterinary equipment consisted of a supply of big green pills and a bicycle tire pump. The pills were administered to any animal—horse, cow, pig, it made no difference what the species or the symptoms were. Animals he treated, with one exception, rarely survived, but it was because as Dutch always said, "You called me too damn late." The one exception was milk fever. The tire pump was used to inflate the udder of a cow suspected of having this trouble. Strangely enough, the afflicted cow almost always promptly recovered. I might add, however, that very few cows actually had milk fever around there since very few were sufficiently high producers to fetch on the trouble.

Later on an educated veterinarian would come down from the county seat and treat milk fever cows with an intravenous injection of a substance with a name about so long. His percentage of survivals were far less than Dutch's.

Anyhow, according to local beliefs, this Hollow-tail was a condition brought on by a grub—usually referred to as a "wolf"—which somehow gained entrance to the upper portion of the cow's tail and proceeded to eat the heart out of it at that point. This caused the cows no end of irritation, loss of appetite, sleeplessness, and mental depression. Loss of appetite did not make too much difference as there was usually hardly anything to eat anyhow. But mental depression in a cow is an awesome thing to behold!

Did you ever study a mentally depressed cow? If not, let me tell you about it. The first symptom you're apt to notice is an unusually rapid flicking of the tail. It is sometimes difficult to determine whether this symptom is occasioned by swishing away flies or by depression. However, if you study the subject closely and find there are no flies you can be fairly sure it's mental depression or estrus. The term "estrus" was never used around Warner as nobody knew what it meant—we all said "bulling". After you're sure flies or bulling is not the cause of the accelerated tail swishing, you should look for the next symptom—difficulty in getting up. Again this is not a sure sign of mental depression. A cow could have a broken leg, a hip out of place or she could be so weak from hunger (malnutrition) she's unable to rise. Incidentally, malnutrition was another term seldom used. Most everybody either said "half starved" or "she's hungrier n'hell."

So now if you've eliminated flies, bulling and plain hunger, you should look for the next and last symptom. Observe the eye closely. If it shows a far-away yearning like wanting some green grass, a male companion, or maybe just a fed-up-with-it-all look, you can be fairly sure it's mental depression. Most anybody can understand how something eating on your tail end could cause severe mental depression, like an over-bearing boss, a nagging wife and like that. I imagine a grub or "wolf" gnawing on the inside of her tail could cause most any cow to have mental depression, or as we used to say around

 Monty Gillian's General Store—go plain loco.

By the way, and this has nothing to do with Hollow-tail, but remembering Monty Gillian's grocery store reminds me of "Skunk" McCrane (not his real name). "Skunk" was an ornery local boy who sometimes worked, but generally didn't. Sometimes he'd have a few dollars in the bank, but usually not. Somehow he never bothered to know his bank balance, but went right on writing small checks to the local merchants without a dime on deposit. One day when he sauntered into Monty's store, Monty hailed him with "Skunk, that three dollar check you gave me a couple of days ago bounced."

"I declare," says Skunk, "now ain't it a helluva bank that ain't got three dollars in it?"

But back to the diseases. There were certain other symptoms for Hollow-tail described by local cow owners such as: "slobbering at the mouth," "moaning and groaning," "blowing up," "running nose," "refusing to give down milk," etc. We had one fellow, Bob H., who was a part-time farmer and part-time shoe cobbler. He was also the best catfish fisherman around, but his knowledge of Hollow-tail and the treatment thereof was known community-wide—even extending to the neighboring communities of Porum and Webbers Falls.

Of course, there were a few like my father and Mr. King, the

 school superintendent, who didn't believe there was any such thing as a Hollow-tail. But the majority who did so believe always called on Bob H. when a cow became puny. Bob almost always responded. He'd treat the cow at no charge provided the owner gave permission for him to fish in the farm tank (pond) if there was one on the place. Otherwise his charge ranged between $1.25 and $1.50 depending on the owner's financial status. Bob always seemed to know about that.

Now about the treatment. The cow was usually "plumb down" so restraining her was no problem. Bob had a special knife which he kept razor sharp just for this purpose. He'd split the upper end of the tail for some five or six inches, sometimes remove a joint or two, fill the vacity with salt, then wrap it tightly with some old rag strips, tie it securely with binder twine, collect his fee, or a promise to fish, and leave.

Of course, the owner, feeling sorry for the cow, would start feeding her two or three times a day with the result that she usually go well. The nonhollow-tail believers naturally always argued that the cow was simply starving to death (hollow-belly) and, consequently, recovered as soon as she got something to eat.

"Hollow-horn" was quite similar in symptoms with one major exception. An affected cow would shake her head like crazy and bawl more instead of the rapid tail-swishing.

In treatment, the medicine—salt—was the same, except that it was considerably more effort to open up the horn than the tail. Bob H. used an old chisel and charged up to $2.00 for the job.

Salt was used also to treat pink-eye, which was referred to only as "watering of the eye". The treatment consisted of rubbing a handful of raw salt roughly onto the eye. Today this treatment would be investigated by the SPCA but strangely enough there was never a blindness and rarely a "blue-eye" in those so treated. Some of the local high school agricultural students argued that a cow receiving this treatment was scared not to get well for fear she'd receive another dose.

Well, I pulled out of Warner at the age of 17 to "go off" to school and since then have not kept up closely with local cow diseases and other community problems. I have been advised, however, that with television and with most people now able to read well these two once dreaded diseases are all but unknown.

The veterinarian is a bright young Vet School graduate who travels around the country in a station wagon or van equipped with a two-way radio and loaded with a hundred different kinds of pills, anti-biotics, sulphas, vaccines in a refrigerator, precision tools and rubber gloves. He uses scientific language with which a good part of the present day farmers are fully acquainted and who use the terms with considerable pride. Back in my boyhood days if someone had used the term "antibiotic" he would likely have been accused of vulgarity or "puttin' on the dog".

On certain occasions it might be said that a sick cow could have some trouble other than Hollow-tail or Hollow-horn, but this was rare with those who believed in the two. Very seldom did anyone treat for lice, ticks, scab, or anything else unless forced to by such things as the Texas fever-tick clean-up.

There was no such thing as a slow or shy breeder among the cows. The skinny old native range cows although seldom weighing over 650 pounds never failed to settle, calving regularly every 10 months, and I never heard of one having calving difficulty. Of course the bulls weren't much larger which no doubt accounts for the no calving difficulties. There was absolutely no such thing as a slow or non-breeder amongst the bulls—they'd positively settle any cow that got in the way, even at a fast trot.

I intend to give these matters further thought.

CHAPTER 8

OKIE CATTLE

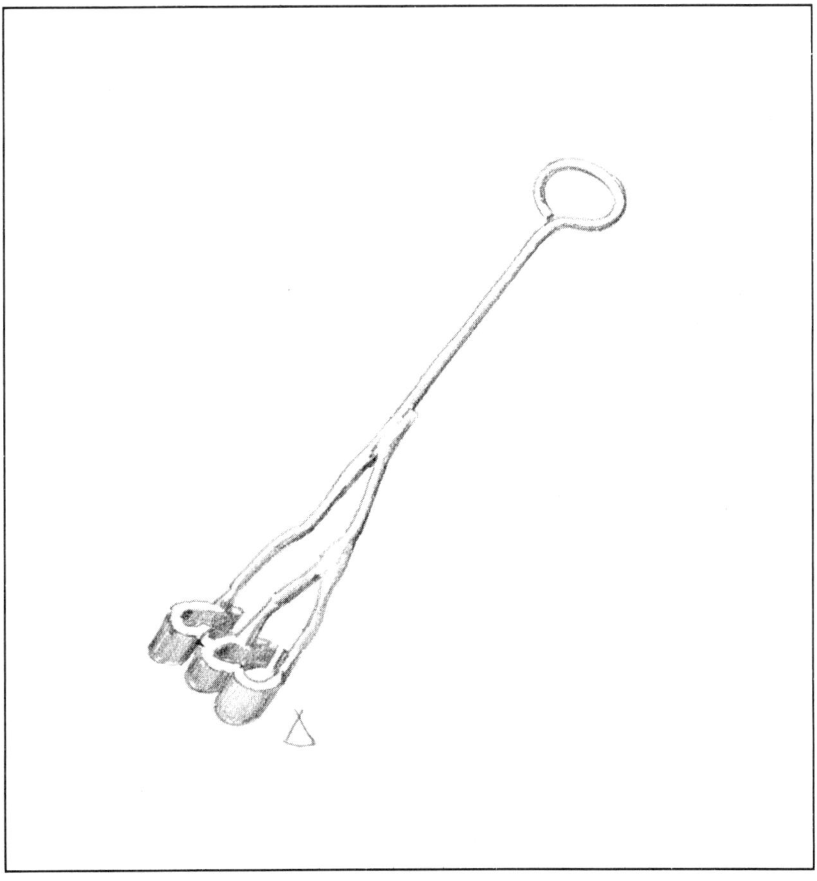

Chapter 8

Okie Cattle

So far as I know, the term "Okie" came into common usage with John Steinbeck's book, "The Grapes of Wrath," which I imagine most Oklahoman's have read and gotten mad about. Just by chance you may not have read the tale, it is based on the dust bowl days of the 1930s. The characters were, for the most part, supposed to have been brought up in the vicinity of Salisaw, Oklahoma. Dust storms, poor land, and the great depression forced these people to load up in anything that would run and head west, winding up mostly in California, where they picked grapes and anything else that needed picking. Their morals would not always bear close scrutiny by the clergy, but neither would a lot of others, in those days, or later.

Not all so-called Okies originated in Oklahoma by any means. People migrated to the West Coast from just about every mid-western state. However, they all got to be known as Okies, like it or not.

Naturally Steinbeck's story is an exaggeration of the actual conditions and happenings, but anyone who lived through those years will well remember that poverty was rampant and near starvation was not unusual. Generally, not always, the migrants were from the poorer land, with much less resources, education and pedigree than those who stayed put. Consequently, the term Okie became associated with the less affluent, sometimes "trashy" element, with poorer or mixed-up breeding. It is no wonder that most Oklahoman's resented the term Okie. I once heard a fine Oklahoma speaker say that, "The only person who dares call an Oklahoman an Okie is another Okie."

Just when the term was first applied to cattle does not seem to be a matter of record. A good guess is that it was concurrent with the great feed-lot expansion in the West and Southwest. In those days, native cattle in Eastern Oklahoma, Arkansas, East Texas and on throughout the Southeastern United States were a far cry from those found in the better herds farther west. The so-called "piney woods" cattle of the area traced to a considerable amount of dairy cattle blood, especially Jersey, with an adequate mixture of Longhorn, Spanish, native stock or just plain cow. They were multi-colored, multi-sized, multi-shaped and multi-tempered.

As the farmers and ranchers of the area gradually became more interested and knowledgeable about the cattle business, brought about by federal allotments for cash crops and land erosion, the Mid-south and Southeast provided an attractive market for well-bred seed stock. The result has produced an amazing improvement of cattle grown in the area. Today, some of the country's finest herds are found east of Texas and Oklahoma and south of Tennessee and Virginia.

In spite of the recent great improvement, many animals still show certain characteristics of the piney woods cattle, especially with their mix-colored hair, their facial features and bone structure.

As with the so-called Okie people, cattle somehow contracted the term Okie also, I suppose for lack of something better to describe them and to differentiate them from cattle with a longer history of "pure" breeding. Why they should not have been called Arkies or Bamas or Crackers or some other term is a bit hard to explain, except as poor breeding was sometimes associated with people in "Grapes of Wrath," or other writings in the dirty '30s.

The term, through constant usage in cattle trading, has become rather standard and well-known, being regularly used in the trade, much to the irritation of many fine Oklahoma cattlemen.

Today, the huge feedlots, especially in the West and Southwest, which have sprung up in the last 15 to 20 years provide an outlet for feeder cattle produced in all sections of America. A large feedlot with capacity of 10, 30, 50 thousand or more must depend on agents to make purchases to fit their feed conditions and their market outlets. Therefore, it becomes a necessity to describe their requirements with some degree of accuracy.

Describing a choice Hereford steer weighing 500-600 pounds is not difficult, and is well-understood by most all in the breeding, buying and feeding business. But appropriately describing a feeder steer whose great-grandmother was a Jersey cow, a grandmother sired by a Brahman bull and himself by a purebred Hereford bull, becomes a bit more difficult. He may be a brindled-red color with a slight hump on his withers and with a white brockled face. Yet he may have good beef conformation and will, with a reasonable feeding period, become a choice slaughter steer. Therefore, in the cattle jargon, he becomes a No. 1 Okie feeder, and buyer and seller alike know exactly what he is when they talk about him on the telephone. In effect, he shows

No. 1 Oakie

No. 2 Oakie

a preponderance of pure beef breeding, but his "cold" background usually shows up in his color and to a lesser extent in his conformation. There is no set pattern to his color. He may be black-bodied and white-faced, black all over, or multi-colored all over. If he happens to have true markings of a breed such as the solid red body with typical white markings of the Hereford or solid black of the Angus, then his conformation will usually classify him. While the term is in common use and well understood by most cattlemen, it is nearly impossible to single out one individual as an Okie of a certain grade.

No. 3 Oakie

No. 4 Oakie

 Therefore, the term is used to describe a group rather than an individual and actually means a "blend."

Commission firms, traders and order buyers will assemble cattle of all breeds and crosses from many farms and ranches in their area and quite often from great distances away. They will then proceed to "blend" into uniform groups according to sex, weight, conformation and quality—ignoring color or breed. These groups become "Okies". A pure Brahman will not normally be used in a high-grading Okie group, but a pure Angus, Hereford or Shorthorn may be, and most such groups will include several that could fit well into a straight English beef breed group.

Oftentimes, those not too familiar with the cattle business will believe that the term "Okie" is synonymous with poor conformation and inferior quality: This is not the case. While Okie groups are normally of mixed colors, with a preponderance of brockled faces and Brahman crosses, their conformation and quality rating runs about the same as with staight English breed groups of the same grade. The term has no reference whatever to where the cattle originate—it is simply used to identify a "kind".

A No. 1 group of Okie feeder steers will just about equal a "middle choice" group of Angus, Herefords or Shorthorns. No. 2 Okies will about equal a "good" grade of straight English breeds—and so on down the lines.

Since practically all Okie groups will be made up (blended) at southern markets, it naturally follows that they will usually contain a fair percentage of Brahman crosses, and in lower grading groups there is apt to be a good percentage of straight Brahman cattle. However, conformation and quality determine the rating rather than the breed.

Some feedlots actually prefer good quality Okies (No. 1's or 2's) over straight English breed groups. They are usually just a bit cheaper, in many cases they carry less fat, often times they are slightly larger framed, they respond well in the feedlot and generally will reach a desired grade (such as choice) when slaughtered. Naturally, their quality response is in a large measure determined by the percentage of pure English breeding, but these breeds rarely get the credit for it. It is my opinion, and that of many capable feeders, that "Okies" without a good percentage of English blood will soon be traded by some other name.

As far as I know, no cattleman anywhere sets out purposely to produce the so-called Okies. These cattle are simply a result of using good sires to "grade up" nondescript cattle. It is my opinion—which a fair number of experiments will support—that even the higher grading Okie mixtures will not perform as well as straight beef breed groups of equal grade. They will not "appear" as uniform, they will not gain as uniformly in the feedlot and the carcasses because of varying backgrounds will not yield as uniformly. Some feedlot operators will debate this opinion. O.K., I'm ready!

CHAPTER 9

I COMMENT ON COWBOYS

Chapter 9

I Comment on Cowboys

General Douglas MacArthur, following his retirement from the Army, once made an impassioned speech in which he said, "Old soldiers never die, they just fade away." Soon thereafter, some wag in talking about cowboys said, "Old cowboys never die, they just smell that way."

The dictionary says a cowboy is "a man usually working on horseback who herds and tends cattle on a ranch." This definition fairly well covers what a cowboy is supposed to be, but it falls short of describing everyone who gets called cowboy.

Certain people will call about anybody they speak to "cowboy." The late Willie Joe Largent, a long-time Hereford breeder from Texas, was typical. Anybody he liked was cowboy. It made no difference where he (sometimes even she) came from, what he looked like, what kind of clothes he wore or whether he'd know the difference between a cow and a nanny goat. If Willie Joe spoke to him at all, he was Cowboy. To him it was the same as Dear Friend or My Fellow American.

Several people are like that.

A real cowboy knows what one is, but since there are probably several million non-real cowboys who may be confused about the matter, and since so far as I know nobody has straightened the thing out for them, perhaps I'd better get at it. Other authorities seem to be too timid to speak up.

While there may be some different kinds of cowboys with which I am unfamiliar, still I do know a batch of them, and if you happen to know others, why you write your own story. Remember, a piece of paper will lie real still and let anybody write anything on it.

If you were to ask me to classify cowboys, which nobody has, I'd divide them up about as follows: I'm doing this all by myself.

There are real cowboys, rodeo cowboys, movie cowboys, drugstore cowboys, dudes and miscellaneous. In the real cowboy group there are (1) working, (2) non-working, (3) foremen and (4) owners—usually called ranchers. A working cowboy fairly well fits Webster's definition with one major exception. He probably spends more time riding in a pick-up truck or jeep than he does riding a horse. In the old days according to the books and a few lies, he lived on beans, hog fat-back and sour dough biscuits. He rolled his own cigarettes, slept on the cold ground, went to town every other Saturday night and chased women for the same reason every body else did. He liked cows, wild women and booze—in that order. But he liked the outdoors, was loyal, hard working, fairly honest and he knew cows' habits by heart. He rode a horse mighty near all day long chasing cows for various reasons, sometimes far into the night. He owned and could use a six-shooter with great skill but hardly ever did—in spite of what the television shows.

Today's working cowboy hardly fits these old yarns except he is also hard working, loyal, honest and he knows all about how a cow thinks. More often than not, he is a family man who lives with his wife and children in an attractive modern house with electricity, indoor plumbing, television—just about every modern convenience that his city cousin has.

He loves and usually rides a good horse on a fine saddle, but doesn't get to do so nearly as much as he'd like. He wears boots on Sunday that cost a pile of money and likely a forty or fifty dollar Stetson hat. He wouldn't be caught dead, however, in a shiny new clean hat. He thinks it is not fitting until it's roughed up a bit so that it looks like it had been slept in. He usually belongs to the local church and he takes part in the community affairs. His children ride the bus to a modern, quite large consolidated school. On the better ranches, the working cowboy earns decent wages enabling him to provide all the necessities for himself and his family. When he is at work, he dresses in tattered western jeans, rough, dirty boots and a battered greasy hat. He is lean, suntanned, clean shaven—a real nice guy.

The non-working cowboy is a lazy shiftless drifter who moves from job to job. There are not many of these left—they've most all moved closer in, let their whiskers and filthy hair grow down to here and joined some kind of a movement. They look and smell like

other hippies. The world would be far better off without them. If decent people had their way, these would all be lined up and sprayed with sheep dip.

Likely as not the working cowboy is named or nicknamed Slim, Shorty, Dode, Doggy, Pip, Tater, or something similar. Never did hear of one named Tobiah, Llewlyn, Ethelbert or Cadwallider—no reflection on these names—they simply don't fit cowboys.

A real cowboy can and does rope, ride any kind of a horse, vaccinate, castrate, brand, dehorn, fix a truck or tractor, cull cows, shoot a coyote, doctor a sick calf, rodeo on Sunday, build a fence, square

dance, take a drink, pray in silence, predict weather or politics, play poker, fight if called on and wither away if cooped up. I never knew a real cowboy I didn't like.

Foremen may or may not be real cowboys. Some come up from the ranks; to do so they have worked hard, learned the business end of the business, pleased the owner and got along with all the other cowboys—but with a firm hand. A foreman can and does keep up with the market trends, knows all about range grasses, nutrition and control of most diseases. He has an air-conditioned office—kept clean by his wife—a two-way radio, keeps records on the ranch, reads all the latest research information and usually makes money.

Of late, many ranch foremen—or managers—are college graduates, even with advanced degrees. I wouldn't want to say which is better—the up from the ranks type or the college trained. It all depends on the guy. The college graduate most often has ranch background anyhow. A college degree usually helps an otherwise good man, but it can surely spoil a kook, once and for all.

People who own range land are usually called ranchers. A rancher doesn't have to be a genuine cowboy, but it helps. Most good ranchers can do anything any other cowboy can plus finding money to operate on and pay the taxes. Today a good many ranches are owned by an absentee—either by inheritance or purchase. Many are women who outlived their husbands or maybe a spinster who couldn't be bothered with a no 'count husband.

Since raising enough money to finance the starting of a ranch or farm is extremely difficult, young men can often get a fine start by managing for an absentee owner. There are a good many successful ranchers who started that way.

I really know of no better way to describe a rancher than to quote a description which appeared in the January, 1968 issue of the *Western Horseman* and written by Larry H. Humphrey:

"Ranchers are usually found where there's cattle-feeding, dehorning, branding, trading, roping, and doctoring. Bankers hate to see them coming, little boys admire them, the Secretary of Agriculture confuses them, city people visit and don't understand them, meals wait on them, other ranchers compete with them, barbed wire cuts them, television glorifies them, but nothing discourages them.

"They like fairs, rodeos, auctions, dogies, hounds, dances, neighbors, forty-dollar boots, Saturday in town, poker, good weather, fist fights and rank horses.

"Ranchers don't care much for poodles, dudes, government men, fixing coffee. They put up with relatives, worms, flies, floods, blizzards, feed salesmen, drought, bad luck, and bad weather.

"Today a rancher must be a salesman, animal nutritionist, vet, biologist, weather prophet, and a banker's calculated risk. He handles more money than most businessmen and makes less clear profit than a paper boy.

"No man is so far from church, yet so close to God. No man gets as much genuine enjoyment out of running water, television, a good game of pool.

"He carries in his pocket at one time: Bull Durham, pocket knife, staples, tally book, one-inch lead pencil, business cards of at least five competing politicians (all of whom he has promised to vote for), cattle ear-tags, fencing pliers, piggin' string, $1.98 watch, billfold (empty), and a currycomb.

"No one gets kicked, run over, stepped on, bruised, cut up, or as mad as he does in a single day's work.

 "He is overly optimistic in: the cattle market, next year, the ten-year-old cow that's never calved, range conditions, the hay crop, and his twice-renewed livestock loan.

"No one is as generous, big-hearted, friendly, dependable, wise or honest, and he will swap anything except his spurs, ropes, or bits.

"He trusts his fellow man.

"The rancher is the producer of meat, the hope of the future, the selfmade man of today. Big business doesn't fear him, the government doesn't subsidize him. He relies on free enterprise and the hope that next year will be as good (or better) as last. He doesn't cry on shoulders when hard times hit, but resolves to do better if he can.

"He is the epitome of the American ideal, and knows that he either must survive without government or perish with it."

The Rodeo Cowboy most often—not always—isn't a cowboy at all, although in the off season for rodeos he may be. Actually rodeos are athletic events, pure entertainment, and the performers are simply

athletes and finely trained good ones, too. Rodeo events normally only faintly resemble real ranch operations but a good performer can and often does alter his style enough to make a top ranch hand. He hardly ever does, though, because he can make much more money in the entertainment rodeo, even though there is no salary—his only earnings come from prize monies. Rodeo rules, event entry fees and all the rest would require a large book so I'll leave that to the rodeo authorities. Rodeos are highly organized events requiring strict regulations and a huge supporting cast. Rodeo cowboys look about the same as any other cowboy—same boots, pants, hat and lean appearance. Many rodeo performers annual earnings reach into six figures and most of these I know anything about own ranches or horse farms of their own.

Of course, there is the rodeo tramp, a sort of hanger-on who barely earns enough to eat on. Therein, the rodeos are not different than

any other athletic event—there are bums in all of them. Top rodeo performers are usually good citizens—fine people.

This brings us to the movie cowboy. Again these are not usually real cowboys—they are actors. When I was a boy and could afford to go to a movie, which wasn't often, I wouldn't be caught in any movie that wasn't a "western". In those days there was no such thing as a "talkie" these having come along in the '20s. The big names were William S. Hart, Hoot Gibson, Tom Mix, Ken Maynard and a few others whose names I haven't bothered to remember. I have no idea whether these fellows could sing or even talk as all they did on the screen was ride hard over the hill to head off the villain at the pass and save the gal who giggled and batted her eyes. W. S. Hart didn't even wear the usual western looking hat—his looked more like a Canadian Mountie's hat except it was dirtier. Hoot, Tom and Ken wore hats quite like those worn today, but like Hart, their costumes were ordinary, without sequins, tassels or other fancy trim.

All this changed with the coming of sound and color. Actors like Gene Autry and Roy Rogers began to pick (or pretend to pick) a guitar, sing lonesome cowboy songs and wear the fanciest duds they could come up with. Today's movie cowboys must be able to talk plainly, own a performing horse or so and above all, sing. They still have to stay on the topside of a horse at full speed across yonder hill. Fist fights and tumbling down a rocky hill are usually taken care of by an extra or stunt man with the hero showing up in the next shot without even a dirty shirt or a bloody nose.

Lately, Broadway actors are cast in cowboy roles, but they seem to experience considerable difficulty in cultivating the proper drawl or sitting on a horse like you're supposed to. Movie directors ought to go down to Albany, Texas and watch 90-year-old Joe Matthews ride a horse. Joe sits there like the real gentleman he is and it gives you a great thrill to watch him.

John Wayne comes about as near looking and acting like a real ranch cowboy as any actor I've watched—not only that, he actually owns cows and is one helluva guy.

One of the mysteries of our time is the drugstore cowboy. Why is a small or large town dude sometimes called a drugstore cowboy? He doesn't even have to wear western type clothes or boots. I have been thinking some of trying to secure a grant to finance a study of this phenomenon. At first this may sound unreasonable, but then I notice where a professor has secured funds to study the effect of different kinds of noises on the hay and oats consumption of sheep. A preliminary report on this subject indicates that sheep eat quite a bit more when soft music is piped into their pens but sonic booms and similar noises scare hell out of them. They gang up in corners, shiver and shake and carry on like a modern dance pair. I imagine if the final conclusion verifies the preliminary report, sheep herders in the west will all carry transistor radios with huge loud speakers tuned to some big city FM station—like Salt Lake City or Deadwood, South Dakota.

Anyhow, a drugstore cowboy likely would not be able to distinguish between a cow and a pregnant orangutan. However this fact does not lessen the usual snide reference to his clothes get-up or his hanging-around-the-corner habits. I've no idea why these dandies are called drugstore cowboys rather than department store, book store or ladies wear cowboys. This needs some research.

Of course, there are the dudes. A dude is described as (1) a man who dresses in a flashy or extremely fastidious manner; dandy, fop, (2) a city person especially one who is vacationing on a ranch. So, a ranch dude or cowboy dude usually comes out from the city on his two weeks vacation, gets a sore rear end from riding a horse on a "dude" ranch and goes back to his city job more worn out than when he came. City kids and college girls like dude ranches and a good many ranches have been saved from foreclosure by earning fees from these city people. When dude ranches were first started, the dudes were supposed to dress in the latest western clothes, stay clean and ride with dignity. It's not that way anymore. Today, the vacationers like it informal, fairly dirty with shabby clothes, the tackier the better. Dude ranches are an excellent place to give city folks, especially the kids, a taste of what the country life is all about . . . . . . I'm fer them!

CHAPTER 10

THE AUCTION STORY

Chapter 10

The Auction Story

The late Bobby Vincent, who was considered an excellent auction sales ringman, once started to write an article about the purebred cattle auction business for his paper—*The Ranchman*. Bobby labored hard and long over the first draft; but when he read it over he decided it "stunk" and promptly threw it in the wastebasket. Right away he started the second draft and upon completion it went to the same wastebasket. The third draft met the same fate. He then gave up the whole project since, as he figured it, about all he could say was, "The ringmen stood around looking like experts, and all the auctioneers tried to imitate Art Thompson."

I expect if Bobby were still around today he'd admit that much more goes on around the auction ring than is covered by his philosophical statement.

Bobby's notion of a good story had considerable merit, so I've done enough research about the auction's history to barely get by. Tracing history is not the purpose of this piece, rather it will be an attempt to enlighten some who may not know about how the purebred livestock auctions operate, a bit about some of the auctioneers, plus a few of the incidents and philosophies of the business.

First a small amount of history. According to Herodotus, for example, auctions existed around 500 B.C. in Babylon when once a year women of marriageable ages were sold on condition that they be wed. Today, some will argue that this was perhaps a cheaper way to obtain a wife than the courtship, church-wedding method now commonly used. At any rate, according to accounts, the most beautiful Babylonian maids attracted spirited bidding and often "above market" prices, while the plainer lasses usually went via a sort of inverse auction system. It worked like this: The auctioneer, instead of asking, "What'll you give for this one?" would start his chant—I suppose they used a chant—by saying, "Would you accept—say twenty dollars—to take this lass home and wed her?" If he got no taker he'd raise his ante up and up until finally some hard-pressed bachelor would claim the miss with enough cash added to pay off the local cafe's past-due food bill. It might be a reasonable conclusion that some of the homelier lasses turned out to be better cooks and more prolific than the dolls.

The experience in selling women very likely provided the basis and the experience for developing the different auction systms used nowadays, namely; the ascending or English system, the descending or Dutch system, and a hybrid of the two. These are the main ones, but there are some minor ones such as the Japanese system which amounts to posting a price for an item or a lot on a board until some bidder raises a finger or gives some other appropriate sign indicating he'll take the item at the posted price. If no finger is raised then the price is lowered and so on until someone does bid. The auctioneer is, of course, the judge of who first raised a finger or nodded. Obviously, a quick finger is important and this may have been one source of the expression "trigger finger".

As to the English or ascending system, it amounts to the auctioneer asking for—usually getting—an opening bid and then calling for a higher amount. At most commercial livestock auctions the sale company has a "starter" who places the opening bid. If no higher bid is forthcoming then the company owns the animal. Occasionally the starter might get carried away with his opening bid being too high, in which case he backs up until someone does come in with a bid. When he does back up he generally does so at a sufficient amount so the company can come out at a future sale or at private treaty—the private treaty sale often being by a standing order. Oftentimes the auctioneer is charged with making the opening bid with the understanding that he backs up when no raise is received.

Naturally this system of "starting" has very properly found its way into the purebred auctions for efficient selling, but any knowledgeable auctioneer will know about where an animal should be started, especially after the sale of a lot or two when he's had time to appraise his crowd. If it's a slow conservative crowd he's apt to open much lower than he calculates the actual worth of a beast, whereas if it's an anxious spirited crowd he'll begin his chant at a higher figure. It's my belief that the system is perfectly honest, and good, for without it an animal worth say $2,000 is apt to be started by some bidder at $100 necessitating an auctioneer hammering away at a snail's pace and boring the buyers. Many purebred buyers are either busy ranchers or farmers or perhaps a businessman who might get up and leave if the sale drags.

Starting too high always reminds me of the late Bill Pace. Those who knew Bill would probably all agree that he was one of the great "characters" who ever practiced the auction profession. He was a southerner who spent most of his life in the southern and mid-south states selling commercial auctions, including mule sales. However, he sold a good many purebred cattle sales, and assisted other auctioneers in a good many more. "Ham" Hamilton relates this story about Bill:

"Bill and I were the auctioneers and it was late in the sale with some very low quality yearling bulls coming into the ring. The sale was being held in an auction market barn. Pace had "started" the bull at $150 and in the usual chant had reached the $170 mark—seeking $180. Realizing he had overshot the mark and while he was taking a moment to see if someone would bid the seventy-five, he was stopped by a brand new ringman. The new ringman was talking to an elderly gentleman who had called him over in an attempt to find out the status of the bidding. He yelled out, 'Colonel Pace, there seems to be some confusion as to just what the bid is.' Pace calmly replied, 'There can't be any confusion because up to now there ain't been no bids—I got a hunnard, 10, 20, Now 30 . . . . '

One time at a sale in Mississippi, Bill started one too high and without too much comment started lower, but was still asking for more than seemed to be offered so he stopped and said, "Now, folks, I have submitted several bids. Won't one of you submit one?"

In purebred sales the opening bid-call and the amount of the raises asked for by the auctioneer are usually determined by three things: (1) The approximate value of the animal, (2) Nerve and know-how of the auctioneer, and (3) Competence of the auctioneer's ring-man help.

As an example, suppose there's an animal worth on the sale day's market about $1,000. A good auctioneer will normally ask for a starting bid at say half the animal's value. Then he'll ask for—perhaps call out—the next high $50 or $100. As the bidding nears his calculated value he will narrow his "asking" to say $10, perhaps $5 or even less, as he approaches the end of the sale. In between the actual figure calls he'll use "fillers" to liven up the chant hoping to keep the sale peppy and the buyers awake. Fillers will vary with individual auctioneers, generally with utterances which mean absolutely nothing. Rhythm of chant, however, in the American livestock auction is of prime importance in maintaining buyers' enthusiasm.

In the Dutch or descending system the opening is higher than the auctioneer believes the item to be worth. If no one claims it at the opening—which they almost never will—the auctioneer keeps lowering until someone does come in. Actually the American auction system is a combination of the English ascending and the Dutch descending—a hybrid. As you can see, our auctioneers will sometimes start too high, then failing to get a taker will come down—descend—until someone does bid, then immediately start back up—ascend—selling out at the last bid—or what he believes to be the last bid.

A recent book by Ralph Cassaday, Jr., a California professor, based on some lengthy research, describes auctioneering as "a unique system of allocating scarce chattels or other property based on price making by competition of buyers for the right of purchase." I think what this means in the case of cattle is that a cow is chattel—most times not scarce however—and price making by competition means bid asking and calling by an auctioneer, usually with more than one bidder in the race until all fall out except the one who then gets the chattel (cow). Once in a while the successful bidder gets cold feet, claiming the auctioneer, or his ringman helper, took him for bidding when he was only waving to Aunt Lucy who just came in or that he was brushing away a face fly in order to avoid pink-eye. In this case an experienced sales force will figure some reason for getting the chattel fetched back to the sales ring and the procedure started all over. An inexperienced force will panic for a few minutes until finally some old pro will rouse himself from an afternoon nap and suggest a way to get going again.

Most everybody knows that there are three main systems of price determination: (1) fixed, (take it or leave it), (2) Private give-and-take or jockeying, and (3) competitive bid. Number three is either by sealed bid, common in the building trade, or by open bidding—auctioneering.

Today, somewhere in the world about everything you can think of is sold via the auction system: airplanes, antiques, art works, cattle, eggs, jewelry, fish, homes, horses and mules, land, tobacco, confiscated police items, vegetables, yachts—you name it, somewhere there is likely to be an auction specializing in it.

In America the auction system arrived with some of the earliest settlers but didn't amount to much until some of the furniture they brought over became worn out and then sold at auction as antiques. Tea, tobacco, cows, and numerous items likewise found new owners through auctions, except these items usually weren't worn out.

No doubt the blackest blotch in our American marketing system was that occasioned by the slave trade in which the auction flourished. Although Dutch merchants are credited with starting this traffic in human beings, Americans were just as guilty in allowing it to mushroom. Books, movies, lectures and millions of words have been written and spoken about the evils of slave trade, so no more needs to be added except that a high percentage of slaves were sold at auction with inhumane practices almost too repulsive to relate.

Today, most auctions are operated with high ethics. The unethical ones soon become well-known and along with catching a "sucker" now and then they soon fade. The sucker is usually ripe for the taking and shouldn't have been at the sale anyhow.

But this piece is supposed to be about the livestock auctions so perhaps we'd best get on with it. Some people are skeptical of auctions arguing that they are apt to get "took" in one. Let's say that there have been abuses in the system just as there are abuses in any other marketing system. However, livestock auctions are as clean or cleaner than any system of marketing and much more so than certain high-powered false advertising for commercial commodities peddled through the business world by the fixed-price system.

Anyone attending the American purebred cattle auction for the first time often views the happenings as utter confusion. For example a visiting Englishman, after attending his first one, gave the following account:

"I had a jolly good time without the slightest notion of what went on. A bloke comes leading this 'haffer' in, the auctioneer chap starts up the damndest clatter—not one word of which I could fathom—and a gang of other chaps standing around right away join in waving books and all yelling yeah! or some such word, at the same time. This goes on for several minutes until finally the auctioneer yells SOLD, which was the first word in the whole procedure that I understood. They lead the bloody 'haffer' out, fetch in another and start the caper all over. I reckon they knew what they were doing but nobody could explain it to me."

Anyhow, no activity in the livestock business creates quite as much excitement and is as fascinating as the purebred auction. For many it is a social occasion with a presale party often lasting longer than is good for the help.

Especially in the last twenty years or so the auction method has become more and more the accepted way of selling purebreds. In recent years among the beef cattle breeds here in the United States there are usually over 3,000 sales in which something like 250,000 animals are sold annually. In addition, it's hard to tell how many animals of other species and breeds are sold at other purebred and commercial auctions. Weekly auctions for all kinds of livestock have expanded into a tremendous business, most having been posted since 1959 until there are now over 2,500 in operation in the United States.

Always the central character in any auction—especially the purebred sales—is the auctioneer. Purebred livestock auctioneers, as in any other profession, will vary considerably in ability and style. However, all the successful ones have similar qualities. To start, a first-rate auctioneer must have a good voice. Good does not necessarily mean low-pitched or high-pitched, although the low-pitched, non-rasping voice is the least tiring to both a sale crowd and the auctioneer himself. "Good" means the voice is strong, durable and rapid or semi-rapid depending on the crowd and local conditions. Above all, a top auctioneer can and does enunciate clearly. No matter how rapid the chant becomes, if the bid-call is clear, buyers can follow with confidence. Of the auctioneers I've known, Ray Sims, the Angus auctioneer, will rank first in ability to best combine speed with clarity.

Clint Tomson, long-time purebred leader and former secretary of the American Shorthorn Association, says this about auctioneers:

"Many of our auctioneers have provided real inspiration for people who had, in many instances, no idea of entering or sometimes returning to farm and ranch programs. I have always felt that Art Thompson did the most in this regard of any man I have known. His appeal was properly based on the responsibility of such programs and the basic support it gave to a growing America. He made you feel proud to be associated with the industry and I think this appeal is still perhaps the most significant in the kind of people we find associated in the business. They are making a great contribution."

Clint continues, "Auctioneers I have known by the dozens, and I respect them for their unique combination of talents and ability to perform a fantastic day's work. They are a combination of actor, statesman, commodity merchant, public relations man, and most of them are still human. If I were to characterize their most essential talents, I would combine thorough knowledge of the product they are selling, a somewhat intangible ability to stimulate and enthuse their buying audience which demands a great sense of humor, timing and, above all, unsurpassed persuasive forcefulness."

A successful purebred auctioneer must be a reasonably good judge of livestock—he must know a REAL good animal from an ordinary one. In today's auction system, there are always sufficient numbers of qualified "ringmen", university personnel, and other experts around to keep the auctioneer posted in case he gets carried away but he can look awfully silly starting a $200 animal at $2,000.

An auctioneer needs to know considerable about the genealogy of the breed he is selling. Here again he can rely on the owner, a breed association representative, ringman or others to keep him posted. Most buyers at today's auctions are fairly knowledgeable about the pedigrees and they'll quickly spot some gross misstatement by the auctioneer. I once heard a green auctioneer say that he reckoned, "Domino is the greatest Hereford bull today." Domino had been dead about fifty years.

An auctioneer ought to be an adept story teller. Nothing louses up an auction like a psuedo funny man who delights in telling a mess of old, worn-out non-humorous stories without one bit of timing to a crowd that's in no mood for even a very funny story. However, absolutely nothing can liven an auction and spur an otherwise lethargic crowd into response like a snappy, appropriate, quickly related story. Better still is a meaningful "quip" just at the right moment. Occasionally some auctioneer will attempt to be funny by using some off-shade story. They never go over well. In ability to use a good "quip" or story to liven proceedings very few auctioneers can equal "Ham" Hamilton.

A good auctioneer must have what, for lack of a better term, we'll call "crowd sensitivity". No two crowds are exactly alike so he must be alert enough to know the difference before the sale starts. At best he will sometimes fail, but the good ones rarely do.

A successful auctioneer instinctively knows when to sell and when to "hang on" for another bid. On second thought, maybe this isn't instinct at all—perhaps it comes through experience, especially after a few mistakes. No doubt the hardest place for an auctioneer to use real good judgment of when to sell is in the consignment sale in which animals being sold belong to various breeders. The auctioneer feels he must hang on even when he knows to sell more quickly would liven the sale and result in more total dollars. Each consignor's interest must be honored, however, so he hangs on for another hoped-for bid. "Draft" sales in which all animals belong to one owner, consequently, are livelier with each animal being sold in much less time.

"Trotting" is a word commonly used to refer to a practice auctioneers use to speed up a sale. So far as I know there's no definition of the term, but here's how it works. Let's say the "opening" bid is $500 on an animal that every knowledgeable person at the sale knows will easily fetch $2,500. Sometimes then the auctioneer will call his next bid at $750 when nobody on the sales seats actually made that bid, or he might call several in-between figures such as, "I got 500—now the 600—600 I got—will you give 700—" and so on up to $750. This is known as "trotting". The practice is used by most good auctioneers and while it is frowned on by some buyers, the same ones will complain if the pace is too slow with the auctioneer having to methodically grind away waiting for each small bid.

It naturally follows that the auctioneer will sometimes get caught with his "pants down". He just simply misjudges the animal's value—or the crowd—resulting in his calling a bid and asking for a higher one when actually the "call" was his, not the buyers. In this case he generally simply stops and does one of several things: (1) He might by

 chance admit he got carried away or, (2) He might find a "fall guy" in one of the ringmen helpers, blaming him for the "mistake" to which the ringman will good-naturedly admit, or (3) He might blame the mistake on someone sound asleep on the back row. On some occasions there is a genuine misunderstanding about the bid so the auctioneer is home free.

There's a story about an auctioneer at a farm sale. It finally came time to sell a pet parrot and, after considerable spirited bidding, the bird was sold. "I suppose the bird talks," the buyer said to the auctioneer. "Talks," replied the auctioneer, "Why he's been bidding against you!"

Clint Tomson told this one:

On one occasion, following a sale at Des Moines, Iowa, we had a buyer who was relatively new in the business. He wore with great pride a Phi Beta Kappa key, and appraised almost anyone who would listen of his new-found knowledge of the cattle business and many moments in which psychology played its part in regard to auction sales. Driving with six "old duffers" that night he illustrated his point with the following conversation.

"Do you remember the Lot 46 heifer which I bought this afternoon?" We assured him we did for many reasons.

He continued, "Do you recall that the auctioneers and fieldmen got into a terrific argument when she reached $550.00 as to who had the bid, and had to drop back to $525.00 to settle the discussion? It was at this moment I recognized that no one really knew what had happened, so I placed a $10.00 bid on the animal and bought her."

There was a long silence during which no one spoke, and finally his chauffeur, who also had the special qualifications and experience of a useful bodyguard, spoke up and, while executing the sharp curves of eastern Iowa, said over his shoulder, "Boss, that reminds me of a crap table I used to play in Chicago at which there were usually six shills and one live one. You was de live one."

You be your own judge as to whether "trotting" is a legitimate practice at the purebred animal sale. It is rather well-known by most livestock auction customers, and it is moreover a source of considerable jesting on sale day. Most buyers accept it, but occasionally you'll find those who absolutely refuse to attend an auction because the system is in use. In my opinion, whatever yours is, the auctioneer should never be accused of dishonesty for trotting—he's simply doing what he believes is best for the sale on that day for most people—buyer and seller alike.

These sale maneuvers can produce some funny happenings. For instance, Darrell Overholt relates an incident when he was working at a calf club sale:

"The auctioneer had 'hung up' a time or two, but had gotten away with it gracefully with a little help from around the ring. Some of the young people at the sale had been doing their own bidding for these club calves, so the auctioneer, partially out of jest, SOLD a calf to a little freckled-face boy about 7 years old who was sitting on the front row beside his mother, who must have weighed at least 240 pounds. As I went to get the boy's name, as innocent as could be, it dawned on his mother that this kid must have held up his hand (the poor kid hadn't) so the mother hit him across the head with a whack you could hear all over the barn as she said, loudly, 'Didn't I tell you to keep your damn hands down?' "

The late Ralph Kuhr told this one:

"At an Angus sale in Sterling, Colorado, when after working hard for a much needed raise in the bid, Francis Dressor turned in 'two and a half.' I felt relieved and promptly sold to his man. When questioned as to who the buyer was, Francis informed me with several adjectives that he did not turn in 'two and a half,' he simply asked if she was 'safe in calf.' "

A George Mercer has written a book called "The English Auctioneer's Manual" in which he describes an auctioneer as, "One who sells property or offers it for sale by a system of increasing or decreasing bids, or by any other competitive methods, or who has on his premises a notice or wording indicating he is an auctioneer." Accepting this description about anyone can become an auctioneer by simply hauling off and putting up a $2 sign in his front yard. Some do just that; however, without doing more they rarely get anything to sell and an auctioneer without something to sell has no more stature than a lameduck congressman.

Somebody else describes an auctioneer as a "functional middleman." In its broadest meaning an auctioneer may perform complete sales service—assembling, cataloging, advertising, sorting, collecting, settling—the whole ball of wax. In the more narrow meaning the auctioneer only "conducts" the sale, leaving all other functions of preparation and settling to the property owner.

In the livestock industry, the auctioneer usually has only the responsibility of conducting the sale. If he assumes most or all of the other duties he becomes "sales manager." I'm not sure who should be considered higher on the pole—sale manager or plain auctioneer. I doubt if it will ever be a national issue.

In the more modern livestock auction personalities, Arthur W. (Art) Thompson was considered by most of the people who make the business click as one of the great ones and to be the most responsible for developing the present system. Since I had known and worked with Art over a good many years prior to his retirement, not long before he died, I had a long visit with him catching up with some of the "good old days" stories.

Even at 82 years of age, Art still recalled much of the development of the present livestock auction market system and the reasons for it. During my visit, I asked why he reckoned the auction method of selling livestock had grown so much and why such a large number of breeders now sell by auction.

"Well," Art welled, "in no other system of selling is there such a great challenge to seller and buyer alike to match wits. The seller (auctioneer) is trying his level best to wheedle the very last penny from the prospective buyer, for the man who hires him. At the same time the buyer makes every attempt to outwit the auctioneer by being coy, even to having someone else do his bidding.

"Then," he went on, "there's always the glamour of sale day—nothing else connected with the livestock industry generates quite the excitement of a well-organized, properly-advertised, expertly-managed auction sale."

Colonel Thompson said he became convinced that there was a great future in the auction business when he was a lad cultivating corn behind a team of horses and a walking cultivator on a hot summer day in a Nebraska corn field. He said he learned the "basics" of the business from a neighbor, at York, Nebraska, by the name of T. W. Smith who sold about all the farm sales in eastern Nebraska.

"I started going to sales with Smith when I was around fourteen years old," Art said. "Mostly I went for the same reason many people go today—just to mix with the folks and hoping to make a bargain purchase now and then. I liked the auction chant and began practicing my own out in the cornfield, behind the barn or any place else I thought no one could hear me. One day I tried out my bid calling on Smith—he liked it, so he said, and promised I could try out at a real sale some day when I was with him.

"That's how I got started. Colonel Smith would sell anything that was worth anything much and let me sell the junk—pitchforks, broken down mowers, milk cans and the like. Many times I couldn't get an opening bid so I started manufacturing my own. Often no one would come in—raise my bid—so I'd back up using some sort of excuse or maybe wind up taking home some no good something—usually at considerably more than the damn thing was worth. Fortunately I didn't get caught too much.

"I wasn't very good at bid calling when I started, but by constant practice I got to be fairly good at it. We had one old cow that I reckon I must have sold 500 times—often to the most prominent people in the country, from presidents on down.

"I soon learned to smell out a free spender from a tightwad. The free spender, of course, was and still is the easiest to sell to; but as any good auctioneer will tell you, we got our biggest thrill from crying 'sold' to some tightwad. I don't imagine things have changed much since I quit.

"In the early days we used to sell horses in farm sales to the traveling horse dealers. These were pretty sharp operators and they knew it. Most of them were what we called 'smart alecks' and I just loved to hang a counterfeit on one of these guys—lordy would they scream."

Again quoting Clint Tomson:

"No special background of experience and training seems to fit any two auctioneers I know, but I do think that well-rounded experience proved to be most useful to any of the more successful individuals. One such illustrious colonel gave me a confidence which I shall keep by making him anonymous. He ran away from home when he was fourteen and became an apprentice rider at some of the county fair race tracks that were plentiful in those days. He became an expert rider and booted home many winners for the owner of the stable, who was described to me as a black-hatted Simon Legree. The only hitch in the program was that the stable owner didn't believe in paying the help, and had a tenacious hold on employment by disallowing funds that would enable a jockey to seek employment elsewhere.

"Apparently jockeys are ingenious like other classes of people, so a plan was devised. As was the custom in some instances, a fast horse might be restrained to the point of a somewhat lower than first-place in order to get the odds down to where the owner might 'chunk' enough money to make a winning purse even greater. His friends decided that the best plan was for him to flash a signal the next time he was told to 'pull' a horse instead of boot him home. Therefore, instead of the prescribed course of action, our hero used the whip from the start while his friends 'pulled' their mounts. The half-mile track took him past Simon Legree twice, and stated at the time he told me the story that he could still hear the hoarse shouts of the purple-faced old man screaming, 'Pull—pull—pull.l.l.l.l!'

"He allowed the horse to continue an extra quarter mile, and, on the back stretch, jumped off, and, at the appointed place, picked up the bundle of cash which more than compensated for the wages he didn't draw, made possible by his friends who had combined to put a big wager down on the nose of the winner. This illustrates the point that, while college is supposed to be the easiest place to gain an education, it isn't the only one."

The late Frank Richards, a long-time secretary of the American Angus Association, told a story which I believe as well as any other illustrates the place and reason for auction sales. Here it is as Frank told it:

"Even in the depths of the depression (many today don't know what we're talking about) the auction method was always a way to bring in some cash—maybe not much—but sometimes the only way to get cash so badly needed.

"Since I know your story will carry much Hereford history, I'm going to tell you about a two-day Hereford sale held in northeast Missouri. The advertising—on a limited budget had been spent, the best auctioneer in the business hired, the catalogs mailed, the tent hired and the cattle made ready. You guessed it—it rained for a week and the night before the sale it snowed. Cattle were knee deep in mud and so were the spectators. Over 300 head were to be auctioned. I think every bank in the county was closed. The sale was scheduled to start at 10:00 a.m. and since you couldn't go anywhere else, the tent was full of people. A few choice females had been selected to open the sale—not that the sale order made much difference on this particular day, but it does help to put your best foot forward giving a little thought to your sale order and the sale arrangements.

"Only ten head had been sold when the first section of the seats collapsed—no injuries. The cattle were not bringing much money, so the auctioneer suggested selling five at a time. The first five brought just as much per head as the singles. After about three groups were sold more seats gave away, so we tried ten at a time—still a steady market so we hiked the number to fifteen at a time. By then all the seats were down—fortunately still no casualties.

"About 3:00 p.m. the auctioneer announced that 'this is the last group of females—we have about 15 bulls and the sale will be over.' We sold one single bull, then three in a lot, then the remaining eleven all at once. At 3:45 p.m. (first day) a two-day dispersion was over.

"The average was less than $100.00, but the selling breeder had CASH—he had created new wealth from his land and feed, he could pay off his obligations to a closed bank and he had saved his farm and kept his family together.

"This and many more similar events again proved to me that the little guy who does his own work, cooperates with his neighbors and supports his breed associations was then, and still is, *the backbone of the livestock business*. Without him the industry couldn't survive and WHY SHOULD IT! The auction ring furnishes his pay day—LONG MAY IT LAST!"

One of the real "characters" in the auction business was popular Fred Reppert who sold many of the purebred sales from about 1906 through the early 40's. Someone described him as "a big fat Dutchman with a voice like a bull in the pampas, a shrill piercing whistle, and about as much 'guts' as anyone in any business is allowed." Most of us around nowadays knew him only in his later years, but there are many fine stories about him—it would take a book to relate them all.

Art Thompson said Fred was the biggest "bluffer" the business has ever known. He could convince a crowd that bids were coming so thick and fast that no one could keep up, when actually no one had peeped. Fred had a reputation of selling to fictitious names then crawling into his car and hi-tailing it away leaving a few head on the owner's place for which he had no use. I never believed quite all the stories for after all he'd go back to the same places as auctioneer year after year.

Back in the depression days when you were lucky to get a $100 bid for a bull, I had a job in Virginia at the time, one part of which was to help promote and sell purebred livestock. Every breeder in the area had a field full of bulls, so in cooperation with a group of them we cooked up a deal to take two truck loads to Swainsboro, Georgia. We hired Fred Reppert to come down and sell them at the local auction barn. The sale was advertised for 1:00 P.M. However, at 1:00 P.M. there was not even a visitor in sight much less a prospective bull buyer. Here we were 600 miles from home, trucking and advertising to be paid for, plus a feed bill and numerous incidentals. Finally about 1:45 P.M. a widow woman and her son showed up, about three or four of what appeared to be cotton farmers and an inebriated cattle dealer. "What'll we do?" I asked Fred. "Hell, we'll start," he said.

We did, with Fred making one of his classic sale opening speeches to which no one listened. Well, sir! We spent the afternoon begging and arguing. We made speech after speech and believe it or not finally sold every last bull in those two truck loads, averaging about $60 per head, from which we had to deduct trucking, feed, advertising and Fred's fee. As I remember, net to the breeders was something like $30 per head. One consignor said, "Anyhow we got them off feed, even though we probably lost at least $50 per head by not going to the butcher with them." About every livestock auctioneer and purebred sale manager has had similar experiences—some worse.

Next day after the Georgia sale we had another for Shorthorns back in Virginia so we drove all night with Fred sitting in the front seat, sound asleep and snoring like a big fat sow. Next morning I saw a very funny sight. Some will remember Will Johnson of the *Shorthorn World.* Like Fred, Will was a rotund man, jovial, likable—everybody's friend. Like Fred, too, he had an enormous belly. Fred and Will were old friends and, of course, got together to visit. I'll never forget seeing them standing there visiting, their bellies touching, but with their faces a full three feet apart!

Pete Peterson, long time fieldman, sales manager, and auction company operator, tells this one about Fred:

"The sale was being held at Amarillo, Texas with Fred the auctioneer. He always insisted on selling the top end of the cattle and when the lower end came in he'd get down in the ring and sometimes would make 'monkeys' out of the boys who were making a living out of the business.

"This day we framed him by bribing a rough talking, mean acting friend of ours to act like he was slightly inebriated. He heckled Fred all through the first half of the sale and then continued his needling after Fred got down in the ring, finally getting plumb mean. Fred challenged the fellow to come down and they'd have it out right there before God and everybody. The fellow jumped up, just as if he couldn't wait. When Fred thought he was really coming he flew out the door on the opposite side so fast that he threw straw clear across the ring. The big guy, of course cold sober, howled, so did everybody else. Fred never forgave us, and shortly thereafter quit coming to Texas."

Earl Gartin was another popular Hereford auctioneer who sold many of the more prominent sales from around 1920 to the early 1950's. Earl had a pleasing personality, he was a fine crowd analyst and enjoyed a wide acquaintance. He was an educated man in the ways of the world, in salesmanship and in numerous other ways, but lacked some formal schooling, especially in the use of correct English. Once when discussing a rival auctioneer, Earl had this serious observation: "You wanta know what his trouble is? He just ain't got no polish."

At many cattle sales there will be someone around taking pictures, most times with a flash bulb attachment. When this happened at one of the sales Earl was conducting he'd invariably remark, "It's lightening! It's a gonna rain!"

Earl developed almost total paralysis of his legs during the last few years he was active. He was a rival of Art Thompson; however, the two were close friends, each having great admiration for the other. They were about the same age. For a few years before he was finally hospitalized, Earl would talk about how Art ought to retire. At a session with fieldmen and others in front of the auction box just before starting a sale, he was holding forth on this subject. "Art ought to quit," he said. "Hell, he don't need the money and he'll kill himself with this travel. Here, boys, help me up in the auction box."

Most every breed will have its favorite current auctioneers. In the Angus breed, one of those long-time favorites is Roy Johnston of Belton, Missouri. This fellow, in addition to auctioneering, is a successful Angus breeder, a fine farmer and about as personable as anybody you've met.

Roy said he started in the auction business when he was a 17-year-old kid in his home county—Macon County, Missouri. Like most auctioneers he began with farm sales—everything from hoe handles and chicken coops to used grain binders.

"I attended the Carpenter School of Auctioneering in Kansas City," he said, "then specialized in farm sales for a year, finally landing a job in the Memphis, Tennessee mule market. I worked this mule sale for 13 years, but as mules gradually gave way to tractors, I concentrated more on cattle. I sold all breeds, including one year on the Denver Hereford Sale. Since R. J. Kinzer talked Art Thompson into selling Herefords exclusively, I decided there was probably more room in Angus and besides the Angus breeders were, and still are, most kind to me.

"When I was selling all breeds," he continued, "I worked with Art Thompson quite a bit—there never was a better one—anyone could learn from Art."

Roy has sold the grand champion steer and other winners at the International Livestock Exposition in Chicago for the past 35 years. The sale itself is a great show and Roy is a tremendous crowd pleaser. He went ahead to say, however, that one of his great thrills was that of selling Kansas City's American Royal Grand Champion Steer—a Hereford—in 1946 at $35.50 per pound to the now-retired Eddie Williams of the Williams Meat Company. This record still stands and may never be broken.

In my visit with Roy I asked about some of his philosophies of the business—about fieldmen, breeders, the future, etc.

"The longer I'm in the business," he said, "the more I realize that an auctioneer has a responsibility to buyer as well as seller. Of course, we're hired by the seller, but—although I know I've been guilty on occasions—if we 'hang 'em on' the buyers continuously, we'll drive 'em away."

We talked about publication fieldmen. "There's lots of good ones around in all breeds. For ours I like fellows such as Mark Dempsey of the *Missouri Ruralist*, Arch Stone of *Wallace Farmer*, and Veryl Jones of the *Angus World*. They may not be flashy, but they're steady, knowledgeable and, so far as I know, honest. Too many ringmen believe their responsibility is to rob every buyer. This kind may create a 'sensation' for a while, but they don't last.

"There's a lot of confusion today about type, size and conformation in all breeds," continued Colonel Johnston. "Looks like we're just bound to go to extremes. We got 'em too little—comprests in Herefords, compacts in Shorthorns, and the same thing—unnamed—in Angus. New we're madly searching for bigness, length—the 'New Look.' I don't want any old 1500-1600 pound cows on my place. They eat too damn much and rarely ever wean a heavier calf than my 1000-1100 pound cows. And, by the way, don't let anybody convince you that we don't need depth and capacity in breeding cattle.

"About the future of the auction business. It's here to stay, no doubt about it. Breeders can do their year's business in one day and our breed, like Herefords, is blessed with some fine salesmen—Ray Sims, Hamilton James, Paul Good, young, alert, personable Curt Rodgers, and a few others."

Every breed—cattle, horses, hogs, sheep—has what might be called its premier sale. Some are consignments while others are private sales. Among purebred Herefords there are several individual breeder sales which attract wide interest, and it would be difficult to single out the one of greatest prominence. However, perhaps the consignment sale which creates the most interest, greatest number of consignor's, most prospective buyers and greatest influence on the breed is the annual National Western Sale held each year in January during the Stock Show in Denver.

To begin with the National Western attracts the largest crowd of purebred and commercial Herefordmen of any event in America of similar nature. They come from everywhere and represent every type, size, range of wealth, nationality and sex. Mostly, though, they are suntanned, lanky and weathered with western boots and big hats. They come to Denver as an annual trek to see and be seen, and to visit with friends and tell tall tales about the much better bull they've got back home.

Auction sale day comes on Tuesday of stock show week and there's an air of excitement from about the previous Friday until the last one is sold, sometimes near midnight on the big sale day.

Herdsmen and caretakers are up just a bit earlier on this important day to groom, brush, curl each consignment to look its best. Sale cattle have been judged for sale order on Sunday morning prior to sale day, with those considered not to be of "Denver caliber" being "weeded" out. The weeded ones can sell privately occasionally bringing higher prices than the sale lots—much to the embarrassment of the weeders.

Herd owners will show their nervousness, some having slept very little the night before. A prospective buyer may be even more nervous. Perhaps he can't decide just which lot will do him the most good in the herd back home or he might have been in a long telephone conversation with his banker about just how far he can go on that "one" bull.

Newspaper fieldmen are up early, too, jockeying for a good position in the sale ring, at the same time making every attempt to be of help to the breed, their publications and their customers. These fieldmen for the most part are a sincere, dedicated group, usually hard workers and hard players. They deserve much credit for their devotion to the business, their employers and to better livestock—more about them later.

Association fieldmen are busy hustling breeders from their territories around the barns to carefully consider a good choice.

At Denver, the sale usually employs three or four of the leading auctioneers and each one is fretting about what he might say to get the high dollar when his "turn" comes.

The eastern cowman is easily spotted with his narrow brimmed hat, his low cut shoes, and his look of slight bewilderment.

Western ranch womenfolk with their tight, but attractive, western pants are bustling around and hurrying to the sale ring to get a good seat. These fine women, many of them consignor's wives, apprehensively watch through the sale of their cattle since a good sale may mean a new television, a car, or perhaps money for sending daughter off to school. It is also a culmination of years of hard work in trying to breed the "right" kind of cattle.

Comes time to start the sale—one of the auctioneers selected to start will nervously grab the mike, clear his throat several times and begin with something like, "Now, folks, we'll soon be ready to start the sale—does everyone have a catalog?" Next he'll probably introduce a few dignitaries in the crowd, call off the publication representatives names, then call on some breed association big wig—usually the president or secretary. The big wig makes a long or short speech depending on his mood, the crowd, and how he's been coached. Short speeches are best but seldom made. He usually goes on at considerable length about what a "great business this is," how the market's "bound to climb," and how his breed "will save the industry." Newcomers may be impressed, but the old heads are either visiting or taking an early nap. Sometimes others are introduced for a "few words" but the few words hardly ever amount to much.

Then the sale starts–this is harvest time for the consignor–his animal represents years of planning, anxiety and hard work. Perhaps his home ranch may have been plagued by prolonged drouth, and adequate financing has been hard to come by. Maybe his banker has been on the phone, the kids need new clothes, the son wants to go to college or start on his own. A good or bad sale on his consignment could mean his keeping the ranch or having foreclosure. Is it any wonder he looks tired, nervous and apprehensive?

But, his is a good beast and buyers are interested. The bidding is lively–and up, up it goes–with our breeder showing a wee trace of a smile. His wife on the third row back unashamedly brushes away a tear or two. She, too, has worked hard–no new dresses this year–gnarled hands and worn boots. But, she comes from families that opened the west and made it great. Ranching is her life and hard as it might be at times she wouldn't trade for Fifth Avenue or a mansion on High Street. These are real people–these cowmen and their wives–sometimes down but never out. And their livelihood comes directly from their land and their cattle.

The consignors are not all hard pressed for money. Some are successful businessmen, in oil, banking, manufacturing, real estate, doctors, etc. They, too, are dedicated to improving the breed and pride shows through when a consignment from their herds attracts spirited bidding and a good price.

Sometimes the sale will be topped by an animal from a little-known small rancher's herd, sometimes from the wealthy man's big outfit. Seldom does one herd supply the sale toppers year after year.

Perhaps the most interesting, most misunderstood and the most welcome person on the purebred auction sale seat is the new farm or ranch owner. Often he's a fellow who was raised on a farm, he may have even despised it when he was growing up and couldn't wait to find a city job where he could get away from the farm "drudgery." So when he finished the local small-town high school he either went away to college where he studied medicine, engineering, law or business, or he found a city job and passed up more formal schooling.

In either case he worked harder and under tensions far worse than if he'd stayed on the farm.

He usually got fat, developed ulcers or a heart murmur and couldn't wait to get back to the farm where he could get away from the city "drudgery."

The newcomer likes the glamour of the purebred livestock business. Most often he's sincere, highly intelligent and will, after the usual starting mistakes, make a fine farmer, a constructive breeder, and a real asset to his community. He explores every source of information and advice, tediously sifts it, applies modern methods, improves his land and has a lot of fun. He rarely makes a profit. Sometimes he's accused of buying a farm and livestock simply to "hide" tax money, especially by the local Internal Revenue Agent. Seldom is this accusation true. Those who doubt this statement simply haven't tried farming.

After all, if the only object is to avoid taxes, who would be stupid enough to battle the elements, gamble on an above-cost market, pay high land and property taxes, buy expensive machinery, pay necessary wages and assume all the other risks, simply to save a few tax dollars? One could make the same investment in municipal bonds, take the earnings and spend the winters in Florida.

I can tell you from experience that anyone in town, a tax office, or any place else who thinks a small-time fellow can come out from town with a few hard-earned dollars and make a killing on a farm—is an ignoramus. It's simply not possible.

Every intelligent patriotic citizen despises a true "tax dodger," but before you conclude that every city man who buys a farm or who retires to one has no other reason than to avoid taxes, you had better have a look at the fellow's accounts.

Instead, we ought to praise those who think enough of the land and the country to risk their life's savings, health and sometimes even a marriage, to save and improve our precious soil. We've already allowed too much to wash away. So to you doctors, lawyers, insurance men, shoe clerks and all others who are interested enough and love the land enough to take all the risks, I say—POWER TO YOU! It is a fact that some farmers and ranchers who have been on the land all their lives have depleted the soil by under fertilizing and over-grazing, and who are slow to adapt to new methods for farming efficiency and improved breeding for better cattle.

Then, too, we can hear complaints about "outsiders" getting into farming in competition with the "family" farmer. I haven't heard this complaint too often from the true family farmer. Most often it's from some vote seeker who thinks this approach is popular with the old-time farm people or by some psuedo leader who thinks the same way. True, there are a few racket operations around which exist on fees from "investors" who are led to believe they can "beat the government" by tax savings. But, these operators seldom last very long simply because they have nothing sound, constructive or even honorable to offer. The ever vigilant IRS man soon ferrets them out. He should.

So, those who believe a retirement or investment farm is a place to "ease up" make a lot of money, or save a ton of taxes better take a closer look.

CHAPTER 11

THE FIELDMAN

Chapter 11

The Fieldman

An integral part of the American Purebred Auction is the help from the fieldmen who are employed by livestock publications. These men are given sundry titles such as field editor, field correspondent, business associates and traveling associates. Among themselves they're known as "road runners." Once in a while you'll hear one called a dirty name but not often. They are hardly ever called traveling salesmen although that's exactly what they are.

Some breed associations own their publications and the fieldmen will double as association and publication representatives. I will take either side of a debate whether this is good or bad.

Most of these fieldmen are fine fellows, each of whom takes his job seriously. As a group, they're hard working and most of the time they know considerable about the livestock business. Traveling from ranch to ranch or farm to farm, helping with hundreds of sales each year, driving most of the night to reach the next place is darn hard work.

Once in a while you'll run into a misfit fieldman with questionable ethics, high pressure tactics, or loose morals. This kind never lasts very long. There is also the beginning fieldman who gets too smart too fast, becomes a pop-off or right away, an authority on any subject. Older fieldmen will quickly straighten him out or if he's unstraightenable, he'll soon hang himself. It doesn't take too much rope in this business.

Ed Huff tells about one such fellow—"This was a publication fieldman who was trying desperately to make an impression on a large and important crowd. He had not been asked to work at the sale, consequently, was sitting on the front row when a cow that was rather loose-from-grass made a couple of turns around the ring swishing her tail. This fellow in his effort to make an impression on everyone, rose up and started to make a speech. Just as he opened his mouth, this old cow made another swish and threw a big wad of manure right smack into his mouth. The quick-witted auctioneer yelled, 'My goodness! If he hadn't had his mouth open, it would have hit him right in the face.' "

Many former fieldmen have gone on to become top auctioneers, field service directors, breed secretaries, outstanding breeders—even governors, and owners of their own businesses.

The publication fieldman's basic job is to sell advertising space for his paper, to write news stories, and sales reports. I have never heard of one winning the Nobel Prize for outstanding literature. I doubt that I ever will. Most of these fellows work on commission, but occasionally one will work on a straight or guaranteed salary. A major requirement for selling is to "service" the advertising customer. Service includes such things as assistance in merchandising surplus breeding stock, locating potential herd sires or female herd additions, interesting new customers in a client's herd and management advice. Since the fieldman travels widely, he is apt to know the sources to solve many breeders' needs. Good fieldmen also demonstrate considerable talent in showing up at coffee or meal time and for bragging on the owner's kids or his wife's cooking. Fieldmen get criticized for favoring the breeder whose advertising account is largest. Maybe this does happen now and then, for after all these fellows are human and they sell advertising for a living. The "wear-well" fieldman tries desperately to avoid favoritism, and those who last the best, do so. Of great importance to a breeder is the fieldman's "ring work" if he has an auction sale to sell his cattle.

For about forty years I've been knowing fieldmen. I've seen very few poor ones and I respect them for their work, their dedication to the industry and for their ability to stand up under the grind day ofter day with long hours in all kinds of weather, irregular eating hours, and dealing with all kinds of personalities.

Many of my most pleasant hours have been spent with fieldmen traveling from one sale to another or in a "bull" session rehashing events of the day or in a "knowing" analysis of the business. Many of the stories, ideas and opinions are unprintable outside of the inner circle.

As in any other profession, fieldmen vary widely in their individualities and their methods but, like auctioneers, there are certain qualities found in all the successful ones. Most striking is their enthusiasm for the business. A good fieldman honestly feels that he has the most important job connected with the livestock industry. To the outsider this trait may sometimes appear to be egotism but it rarely is. A good fieldman knows when to keep his mouth shut. A pop-off soon pops out.

Since ring work has become such an important part of a publication fieldman's job, it is interesting to hear some of the appraisals contributed by auctioneers, fieldmen themselves, and others:

. . . "Becoming a good salesman perhaps most of all makes a good ringman. He needs to be a clown at times, serious consultant at other times, but always aware that the principal reason for being associated with a sale is collecting money for the man for whom he is working that day. I really don't believe a good ringman can be described. He either is a good one or not, experience simply makes it easier for him —he already has the ability when he steps up to his first sale ring.

"One of the greatest I've known is O. R. (Pete) Peterson. Pete had the uncanny ability to sense the right move—the right moment for a bit of clowning and I think he understood cowmen and their moves better than anyone I've known. Pete didn't contribute any glamour to a sale from the standpoint of beauty but he could damn sure contribute some glamour through confusion."

Here is another ringman's appraisal . . .

"A ringman must be personable, full of wit, quick thinking, mind reader of sorts, politician, and most certainly have some ability as a modern judge of the breed he happens to be working with."

And . . .

"A ringman, to be good, must be able to instill confidence of the buyers while at the same time assisting to create an atmosphere of enthusiasm that will cause people to want to buy . . . to participate in the action."

Also . . .

"I know not all will agree with me. It isn't the man with the high boots, big hat and loud voice that shouts and screams everytime someone else gets a bid. I think a good ringman is a fellow who keeps his mind on the sale, knows where the bids are coming from and above all lets the auctioneer in charge be the leader."

Still another . . .

"A good ringman knows how to work as a part of the team on sale day. He keeps the confidence of the bidders and knows where the ball is. The loudest show-off usually isn't the most help."

And . . .

"A good ringman is (1) always willing to accept the challenge of salesmanship, (2) he can think quickly, (3) he understands sales psychology and instills confidence that his efforts are in the best interest of both buyer and seller, (4) with a good auctioneer his thoughts are so parallel that the two can communicate rapidly and effectively without uttering a word and (5) to the buyer, he represents honesty and integrity thus giving the buyer confidence. Some ringmen are lacking in (3) and (5), consequently, their effectiveness dwindles as time goes on."

And finally . . .

"All men cannot be good ringmen and we certainly have some around now that just can't fit. I don't know that I can describe one but after he has been there awhile I am positive that I can tell whether or not he will make it. L. R. looked like a flat failure the first time around but he developed what I shall call 'sale savvy' and made it. Some boys just can't do it and a good many of those are working now.

"Sale savvy tells a man where the last bid is and how much it is. This is most difficult with all the chatter taking place. It tells him that the best prospect in his section is the man that has already bid one or more times on the animal selling—although he can't devote all his time to this man and fail to pick up a new man. Sale savvy will tell him whether or not a prospective buyer is the kind that you need to talk with in getting him to bid or whether he must be left alone—whether you can 'kid with him' and get him in or whether you must be 100 percent business in your contact."

▽

CHAPTER 12

TALES FROM THE ROAD

Chapter 12

Tales from the Road

Most livestock publication and breed representatives work hard. They are dedicated to their jobs, they love the glamour of the business and the people in it. They also have a lot of fun along the way with a good many of the incidents worth relating. Here are a few examples:

Curt Rogers, the auctioneer, tells about Hobe Hunter. Hobe was manager of an Angus operation, The Twin Valley Ranch in California, owned by the actor Fred MacMurray. Curt says Hobe tended to be on the playboy side and without a doubt the greatest non-professional entertainer "I ever met." At one time he was a fieldman for Clint Tomson with the Shorthorn Association. One story which Clint could relate better is when he had sent Hobe to a series of field days in the Dallas area. After the third day, which was five days after Hobe had left Chicago for Dallas, he had not shown up in Dallas. Clint became concerned and finally tracked him down at a "friend's" in Minneapolis. The conversation on the telephone went something like this:

"Hobe"

"Yes"

"Where are you?"

"Oh, just a little north of Dallas."

A wild or unruly animal can be extremely dangerous at an auction, especially if the sale ring or other arrangements happen to be poor or flimsy. However, barring injury to someone you'll see some funny incidents. Here are some as told by Tom Sherlock, Dick Kane, Clint

Tomson, Joe Gossard, Merv Aegerter, Charlie Corkle and Pete Swaffar.

Clint Tomson:

Aside from the always serious part of auction sales, that of bringing in enough cash to keep a program running, the lighter moments provide stories that are legend, with a naturalness of funny things that happen that could fill volumes if reported by the dozens capable of recounting such episodes.

A wonderful Shorthorn cattle breeder named Antoine Hekrdle of Stanton, Nebraska, comes out in focus as providing the most sale-stopping action I can recall. At a sale in Norfolk, Nebraska, he offered twenty-two short two-year-old bulls and twenty-two two and three-year-old cows which he had trucked to the spacious community sales barn which handled thousands of feeder cattle brought in there for disbursement to that area. Time had not allowed the Hekrdle family much opportunity to halter-break the bulls, and this well-recognized condition was ingeniously compensated for by the halters made out of one and one-quarter inch "hayrope" that had served its usefulness in bringing big loads into the mow. In order to be sure that no animal escaped, a good eighteen feet of generous length adorned each halter, and the tableau was then established on the following basis: The sale ring door would slide back, the rugged Antoine grasped the end of the halter rope securely, and, hand over hand, began reeling in the slack. The bull was forcibly made to enter the ring through the efforts of four to six stalwart assistants from the rear. A four-footed furrow indicated clearly that there were other places he preferred to the sale ring, but, upon reaching the doorway, they usually made one big jump toward the center of the ring.

At this point, Antoine would gracefully, and with complete confidence that the job was well done, throw the eighteen feet of rope over the back of the bull which was then allowed to make his own way around the ring. He would then make a knowledgeable and persuasive speech concerning the respective merits of this particular bull, spiced with sometimes repetitious remarks about the quality of its ancestry.

This tableau was repeated in an almost identical pattern so that the crowd would seemingly strain with Antoine in his hand-over-hand efforts, and their audible chuckles were a part of the entry of every beast, until the first cow with a two-week-old calf became the candidate at the end of a halter rope. Their chuckles gave way to the sounds that always accompany obvious mayhem as this particular female saw Antoine as her target and required no pushing, no pulling, just a well-coordinated charge. Upon watching Antoine's expression,

which, when the taut long rope he had been handling became a limp omen of what was to occur, he reacted like the athlete he was, and, with one surge raced across the wide ring with the cow tickling his trousers with her horns with repeated thrusts. Nature must supply either superlative reflexes or an extra dose of adrenalin in such circumstances because he literally made the longest broad/high jump I have ever witnessed where I, as a ringman, was able to reach over an eight-foot railing and assist him bodily to a seat on the railing, with the cow still threatening underneath.

The screams of the crowd gave way to full enjoyment which didn't dissipate for a full two or three minutes and, when they did, Antoine told them solicitously that she was the most gentle cow on the place and wouldn't hurt anyone. This broke them up again for

a full twenty minutes, during which they repeatedly wiped their eyes, rolled on the floor and groped for air until Colonel Jack Halsey could get them back to order. It was a tableau we will never forget.

Clint Tomson:

An event of note was held at a sale sponsored by the well-known Dave Canning. Recognizing the value of pagentry, he had programmed three different entries for the bulls which characterized their backgrounds. For the Scottish champion he had four pipers in full regalia enter the ring before the bull. For the American champion he had the Stanton Military School Fife and Drum Corps glorify the entry. But the best was the Virginia breed champion for whom he had a lovely lady with a six-foot hunter's horn depicting the environment which characterizes the hunt.

Unskilled in beef cattle, she was to take her cue from those of us who promised to let her know when she should sound off. Although a lovely lady, she had felt the need of stimulation to keep up her courage and, as the bourbon bottle from which she sipped from time to time went down her spirits were raised and her relaxation was apparent to us all. We were told that the bull would follow a group of heifers, and her scrutiny to determine whether the bull was indeed in line involved a hands-and-knees approach that everyone in the crowd noticed, but of course did not understand why. Her excitement at finally detecting the animal that was the bull was beautiful to behold. On cue, she then raised the long horn to her lips and blew—and blew—and blew-w-w-w, but unfortunately no sound emerged. The crowd, fully appreciative of her efforts, simply couldn't contain themselves, and Manager Canning, hopeful that he could spare her further embarrassment, kept saying, "That's all right, that's all right," but at this time her numb lips pursed sufficiently to create sound, and she blew for several minutes, with everyone trying to stop her to no avail.

Two ringmen, recognizing true pandemonium, dashed in and out of the arena baying like foxhounds, which only served to weaken the crowd's resistance a little more. It was so funny that three of my good friends, later that evening on a plane bound for Kansas City, were told to leave unless they could control themselves.

Merv Aegerter:

A Shorthorn sale was being held in the area of Pawnee, Oklahoma, and the auctioneer that day was Colonel Roy Pauli, of Broken Arrow, Oklahoma. The sale was being held in a little lean-to next to the barn, and on the back wall they had placed four bales of straw. They had driven two two-by-fours into the ground in front of this, and nailed a one-by-twelve board about four feet long across the top of these. This was where Roy Pauli was working. With a man of normal height the top of the stand would have caught him just a little bit above his belt, but because Roy was very short the top of the auction block caught him just a little bit below his chin. So, as he worked, there was quite a bit of Mr. Pauli still protruding in front of the footboard which served as his workbench.

The sale started with bulls whose horns had not been trained and the bulls were not too used to the halter rope. As each one was presented in the ring, they bounded out of the barn door, with one of the big tall brothers at the end of the 20-foot piece of rope. He then would anchor the bull in the center of the ring and let him trot around in front of the crowd.

Colonel Roy got along real well on the first two bulls and when the third one hit the ring his courage was hitting its peak so he started this bull at the selling price of the last one. On this day I'm sure some of the real top auctioneers would have asked for a bid of probably $10 or $25 more, but Roy asked for an extra $50, and to his surprise, and everyone else's, he got it. By this time, of course, the ringman was shouting rather loudly, and Roy's chant was booming. This big two-year-old bull at first was a little bit shook up by all the commotion, but as he pranced around the ring at the end of this 20-foot rope, making the big circle, he finally decided that most of the hullabaloo was being created by this little short fat fellow on the block. On his next swing-around he just flipped his head as he went by Colonel Roy and unzipped his trousers from top to bottom. Colonel Roy wasn't to be outdone, so he turned his back to the crowd to zip up his pants but kept his face to the crowd calling for that next hundred dollars. All the time his face kept getting redder, because his shirt-tail had gotten caught in the zipper and when he finally turned around the zipper was about two-thirds of the way up, with a piece of shirt-tail hanging out. He was quick witted enough to sell the bull immediately but, of course, as you can well imagine, this broke up the sale for about twenty minutes. In the meantime, Colonel Roy called for a couple of boards to be tacked in the front of the two-by-fours, not only to hide his busted zipper, but to keep him out of range of further incidents of this nature.

Joe Gossard:

This happened at the fairgrounds, Newton, Illinois, at a sale being held for "Ham" James. All was progressing nicely when a heifer broke out of the incoming chute. She ran past the loud speakers, which were on towering stands, past portable bottled gas heaters, between the sale ring and the crowd, cut out at the corner and went around behind the bleachers. At this place, the ladies had set up tables in a square and were serving from inside this square. The heifer jumped over the tables into this square with the ladies, then jumped out again over the opposite side and ran between the bleachers and the wall, where she became trapped and some men lassoed and haltered her. During this entire spree she didn't even so much as spill a cup of coffee. No one was hurt or even bruised, no tables upset, no damage done by the heifer. One person got caught in the wire leading to the loud speakers, but another person noticed the broken speaker wire and spliced it. The entire action took less than thirty seconds and the sale went on as though nothing had happened.

Dick Kane:

During a dispersion sale for a South Dakota breeder, the sale ended with a group of young bulls, 10 to 12 months old, which were selling one at a time in the ring—loose. One bull got a little excited and dashed into the ring just as Al Hamann was turning around with the previous buyer's name. The bull didn't seem to approve of Al's looks, so took after him. Al, being considerably over his best fighting weight, ended up down in the middle of the ring half covered with straw, but not before the young bull had apparently hooked a horn in his belt and jerked his pants down. When they got the bull distracted and out the door and we were assured Al had not been hurt, one of the funniest sights that I have seen at an auction was Al standing there in the center of the ring and reaching down to his knees and pulling his trousers back on.

Tom Sherlock:

At an Iowa Angus sale, the man having the sale had about forty big cows with calves. There was a large crowd all around a high fence in a big yard adjoining the barn, where the cows were confined. One at a time they were let out to be sold. About the fifth one out was a big wild-eyed snorter and the show was on. She made a complete swing around the yard and then headed for the high board fence on the other side. The late Milo Sheets was innocently standing in the enclosure, so this cow decided all of a sudden she didn't care for Milo and took after him. Although Milo was never known as a 9:5 second man, I'm sure he was at least that fast in this instance. So was the cow. They reached the fence in a dead heat with Milo leaping for the top rail while the cow went underneath taking the fence with her. Milo, fortunately unhurt, was left somewhat suspended trying to climb the fence that was no longer there.

Charlie Corkle:

In a bull sale at Deshler, Nebraska, a bull became excited when led into the sale ring, took to his leader and not only pinned him to the side of the ring, but was really mauling him. Someone warned the ringmen who naturally were facing the folks on the seats. Gary McDonald, who was on the opposite side, ran across, planted a boot to the rear end of the bull with the force of our greatest football punters. The idea being to divert the attention of the bull—it did, too, for the bull likely carried the boot mark the rest of his life. At any rate, the bull turned and started for Gary but before the bull reached him he was back across the ring and seated on top of a post as unconcerned as though nothing had happened. The leader had been lifted up out of the ring not too much the worse for the battering.

Western ranchers and cattlemen are usually tall, broad, robust and weathered. Herb Chandler, the great story teller and poetry reciter of Baker, Oregon, tells about the time a bunch of cowmen including himself entered an elevator in Baltimore when they were on a meat packing tour. In the elevator when they entered, were two short, slightly built gentlemen dressed in the proper eastern businessman's attire. Of course, the westerners towered over the two. After looking the cattlemen over, one little man remarked to the other, "Just look at 'em. Big, rugged beef eaters—all of them—and look at us! Little shriveled up clam eating sons-a-bitches."

It is pretty generally known that at some high-powered sales a little "giggle juice" is sometimes served ahead of time to loosen up the crowd. A dispersion sale in Oklahoma started at 11:00 a.m. and, without stopping, at 1:00 a.m. the following morning they were still selling. The owner was in the auction box and Pete Peterson who was helping with the sale remarked to him that, "This is the longest continuous sale in my memory."

The owner said, "Let's celebrate. I'll get a case of booze and we'll give everybody a drink." This he did and we told everybody who wanted a drink to come and get it. Everybody came. Soon afterward, we finished the purebred cattle and started on the nurse cows, dairy heifers and calves.

"You never saw such a dairy cow sale," Pete relates, "D. J. from Oklahoma started buying the Holstein heifer calves and wound up with a semi-trailer load. He had no use for them, so the next time I saw him I asked why he bought so many. He said, 'When I unloaded them I was surprised that they were calves. I thought I was buying yearlings'."

One time a bunch of us were on our way to a Hereford event in Chihuahua City, Mexico. Pete Peterson, Jack Turner, Billy Bob Watt and several others were along. We stopped overnight in Odessa, Texas. Before supper, Pete had a jug with most of the crowd imbibing as it passed around. One fellow, whom you all know, came down a little late to eat so somebody inquired as to where he'd been. His thick-tongued reply was, "I've been up having some of Wheterson's Piskey."

One year at the Denver Show on Sunday morning, the sale bulls were being judged for sale order. A certain Kansas breeder had a bull of which he was very proud, but which wound up near the bottom of his sale class.

As the fellow left the judging ring leading his badly beaten bull, a friend asked, "How'd you come out?"

"Not too bad, I guess," he replied. "Sure took a hell of a lot of 'em to beat me."

One time, a good many years ago, old timer Herb Chandler was judging Herefords at Fort Worth. A long time and close friend of Herb's from Texas was showing a bull calf he had previously sold to another Texas breeder at a long price. Herb knew nothing about the transaction. He had lined up his class and appeared to be about finished, with this particular calf standing far down in the class. Someone knowing about the deal stopped Herb with, "Do you know what that calf is you've got standing way down there?"

"No," says Herb. "What is he?"

"Why that's the calf that____sold to____at $____."

"Did he get the money?" asks Herb.

"Damn right he got it."

"Then leave the son-of-a-bitch where he belongs."

Several years ago in a farm visit with what was then considered a top Eastern breeder, a group of us, including several newspapermen, were being shown through the main barn. Two of the young herd bulls were in individual box stalls. The herdsman led the first one out in the alley and the owner proceeded to give a lengthy lecture on the perfectness of this bull.

"Ideal head," he said, "A middle that simply can't be beat, and just look at that hindquarter—I doubt if there's a bull in America as good. In fact, I can find absolutely no fault in this bull," and on, and on, and on.

The bull was put back in his stall and the second one led out.

"Now here," says the owner, "is a bull that we consider quite a bit better than the other one."

Being proud is one thing, but a big line of gobbledygook seldom amounts to good salesmanship.

Of course, I suppose a breeder can be just a bit too "honest". For instance, in a visit with a big likable guy, some of us were watching his yearling bulls pass by. It was a better than average group—very useful to commercial cattlemen. The owner spent most of his time pointing out the faults of each without much mention of their good qualities—of which there were quite a few. Finally one of the plainer ones went by. He was by no means a cull, but hardly up with the others.

"Now there," says Henry, the owner, "is a son-of-a-bitch and there are also some other things wrong with him."

This happened in a sale at Tulsa, Oklahoma. A lady waved her hand and the auctioneer immediately cried "sold" at something like $2,000. "No! No!" she yells, "I was simply waving to my friend across the ring." So they start over and sell the animal to another lady who it turned out was the one who was simply waving back to the first one.

When asked about his greatest sale flop one well-known auctioneer, George Morse, replied: "Now, as to the greatest flop I ever had, I'm not sure you will want to write this, but it was the sale of cattle whose pedigrees were questionable and they were selling mostly calves. I will have to admit the calves looked good—they were bloomy, but the crowd was small and the day was very cold. I was handed a marked catalog which indicated about five of the calves should bring a fairly high price and, of course, we did not get those sold. However, on the seats were two well-dressed men, one sitting on the left bleachers and the other on the right, with their attractive wives. They bought about 90 percent of the offering and I left there feeling that the only cattle that didn't get sold were the two or three of those marked in my catalog and that we had gotten a tremendous job done selling the remainder of the cattle to the two 'newcomers who were just founding a new Hereford breeding operation'. It was six months later that I learned they were only protecting the owner, and the best estimate I can make is that we got approximately ten head of cattle sold."

The late Bobby Vincent was regarded by all who knew him as one of the greatest "characters" around the cattle business. "Ham" Hamilton tells about the time at a sale when Bobby was having a little difficulty collecting his thoughts—which was unusual. He wanted to make a speech about a certain animal so asked the auctioneer to be recognized. Just before he could start his comments the bull in the ring let out a mighty bellow.

"Tha's jus' what I wuz gonna say," says Bobby.

Claude Willett tells about the time he and Foss Palmer passed through the small town of Foyle, Oklahoma, on their way to a Turner Ranch sale at Sulphur. Claude remarked to Foss that Bobby Vincent was born at Foyle. Later in the day they saw Bobby at the sale and Foss said to him, "Bobby, we came through Foyle this morning and Claude tells me that's your home town."

"Yep," says Bobby. "I was born all up and down that country."

Charles Corkle of Norfolk, Nebraska, has been connected with the auction business for longer than most of us can remember. He's one of the nicest guys anyone could know. Although not blessed with the strongest voice in the business he more than makes up for this by his "savvy," friendly nature and wisdom.

Here are just a few of his gems:

In the past, the purebred auctions, the establishment of prices by competitive bidding, have served to set the value not only on the livestock going through the auction, but to a very high degree the value on all the seed stock of their respective breeds selling privately. I realize that we have seen many changes in agriculture and in the livestock business. Progress necessitates change but I doubt if all change necessarily means progress! Our marketing system has been undergoing many changes. Many changes have taken place in agriculture and no doubt many more are to come. Some of our long-range forecasters see the family farm and ranch taken over by giant corporations. I never attend a 4H or Vocational Ag show or sale but I think of what a sad day that would be for America if this were to take place.

Comparatively small herds have in the past produced many of our more influential sires—maybe not because of their lack of herd size but rather because it was necessary for them to use the very best of the then less popular family lines and as a result were in a position to furnish a needed new blood line to further improvement. Every breed of livestock needs the development of every family line and I am thinking of the sire side which shows any promise at all. No doubt we will always encounter from time to time some type of genetic problem. There have been times when if artificial insemination had been in general practice to an extremely high degree an entire breed could have been ruined.

George McGinley, Sr., a Nebraska ranchman took me by the arm after a sale in which I had worked the ring for an auctioneer who had put on quite a demonstration of attempting to be funny. He said, "Don't you ever do that—an auction is the most important day for the seller—it is really a business day for him and for the man who is buying—a little humor and wit is quite all right, but nothing surpasses the reasoning of what an animal can do for the prospective buyer and just a good smile."

My Dad, who had been in the farm implement business in addition to his farming operations until his sons became old enough so he thought the farm was the place for them to grow up, told me when I started in the advertising business that I would really have to believe in whatever I was selling if I were to be a success—that nothing was ever gained by criticizing a competitor or a competitive product.

Changes have come in auctions as well as in most every other line of endeavor—farmers and ranchmen are more familiar with this method of selling and buying. The purebred auction of today is attended by comparatively few who have not either sold or made purchases in auctions. The press, TV and radio keep us all well informed.

Perhaps the outstanding auctioneers with whom I first worked and who will surely go down in auction history as all time great salesmen would make some changes in their opening remarks and their reasoning during the sale.

They would still hold the attention of those at the ringside by reminding them that good livestock is the basis of better agriculture. They would continue to hold their interest by stating the changes that have come and the present-day goals of that particular breed.

Too, I think they would be successful in creating the desire of many to improve their own position in the farming and ranching business by producing a better quality product—the quality that the majority of our over 200,000,000 Americans prefer.

Pete Swaffar:

Once in a while cattle owners let the glamour, excitement and promotion fever run away with their enthusiasm.

This happened in Ohio. The breeder, now out of the business, was a victim of his own exuberance and presale excitement. He had previously held some of the Shorthorn breed's best auction sales and was frequently suspected of being acquainted with all selling gimmicks known. He also invented a good many. In his speech at the opening of a sale, he'd gone into extra detail of how if the cattle weren't satisfactory to let him know, that the cattle would perform extraordinarily well, his belief in them, etc., and on, and on, and then in an effort to say the sale was commencing, said quite proudly,

"____and now we're ready to give you the works!"

Suddenly realizing what he'd said, he was speechless, and perhaps just as well—any explanation would never have been plausible.

Sitting room around most livestock sale rings is usually limited, making it necessary to use steep bleachers. Sometimes women will be seated high up on the bleachers and not be too careful to keep skirts down and knees close together. Following a sale one time when this happened one of the ringmen wrote the following poem:

A bunch of the boys were whooping it up at a sale
While the guy at the mike was telling a tale

And high on the seats a lady in green
Presented a view that was a sight to be seen

So the boys gathered around an old bale of hay
They giggled and gazed and some pointed that way

All the while the gal in green was flappin' her jaw
With men and bulls all pawing the straw

Finally a fellow just to relieve the stress
Yelled, "Lady in the name of goodness, please lower your dress."

At one time, each summer, employees of the American Hereford Association held a picnic bringing along members of their families—and the unmarried girls bringing along their boy friends.

One year the AHA Board of Directors was in session at the time of the scheduled picnic, so Jack Turner invited the Board members to attend the picnic. Most of them did so. Among the members were Herb Chandler, Oliver Wallop and Doug Gay.

The picnic was held at Wyandotte Lake, some 30 miles from the downtown hotel in Kansas City where the members were staying. Jack drove one car with Herb, Oliver and Doug as passengers. Other Board members and employees were in different cars.

The picnic lasted until around 10:30 p.m. Just before breaking up, Doug and Oliver went out back for a little relief not returning hurriedly as they were visiting about Association affairs. Jack and Herb decided it was time to return to the hotel. They looked and looked around for Oliver and Doug, finally deciding they must have returned with someone else as the others had all gone home. Consequently, they returned to the hotel and immediately went to bed without thinking further about the matter.

In the meantime, Oliver and Doug returned to the picnic site and found everyone gone. Here they were, 30 miles in the country, no telephone near and park traffic all gone. It was some 5 or 6 miles to the nearest telephone but, of course, being strangers they had no idea where one might be. So they set out on foot and after a hike of some 5 miles they were picked up by the park police. They told the cop their story, and here was his comment:

"Now I've heard it all! I've picked up a few girls who'd been kicked out to walk home but you two are the first men."

Anyhow, he delivered the fellows to the nearest telephone from where they were able to get a cab back to the hotel.

Next morning they naturally jumped Jack and Herb for leaving them stranded.

Herb's reply was, "I saw your hats on the back seat, but it never occurred to me that you weren't under them."

The story got around that Oliver and Doug were actually the victims of the old, old "Snipe Hunt" gag. There now hangs on the AHA Secretary's wall two moth-eaten snipes under glass, below which is a beautiful brass plate inscribed as follows:

These Snipe Caught In
Wyandotte Park, Kansas
On the Night of Aug. 8, 1953 by Directors
American Hereford Association
Douglas Gay and Oliver M. Wallop

At a weekly auction barn in Mississippi I stopped in to catch up on how things were going. They were selling razor-back hogs–pigs, shoats and little old scrawny sows. One of the latter came moseying into the ring with two equally scrawny pigs following. The local auctioneer leaned out over the box and yelled:

"Get on now boys, here's one with twins!"

At Tazewell, Virginia's, weekly auction they were selling milk cows. Cow traders sometimes switch calves on milk cows–putting either a better or poorer calf on the cow being sold, depending on the calf market prices. On this particular occasion a medium-sized Holstein cow came in with what appeared to be a Jersey-Holstein calf and followed by the cow's owner. A prospective buyer on the back seat yelled to the owner, "Is that that cow's calf?" Without hesitation, the owner yelled back, "Damn right, it's hers. I bought it yesterday and gave it to her!"

The prospective buyer still appeared interested so he shouted another question. "How much milk'll she give?" Again a quick answer, "She'll give a bucketfull twice a day." He didn't bother to indicate what size bucket.

Very seldom does it pay to mix other business with a livestock auction—even religion. The late Jack Halsey told about an experience at a sale in Mississippi which he and Art Thompson were conducting:

"We were selling for two brothers who were 'Holy Rollers' and they had a promotor there who was trying to raise money for the church. The brothers wanted him to make a talk at the opening of the sale. We tried to impress on them that a religious event at sale time would not help the sale. The brothers, however, insisted that if they could save one soul it would be worth it. However, so far as was known, the fellow didn't save a soul, raised no money, and the sale was a complete flop."

Another Jack Halsey story about a sale in South Dakota:

"The 'sales force' had gone into the owner's house for lunch before time for the sale to start. This was a dispersion so after lunch we all hurried out to the sale barn. The owner was due out to 'open' the sale. We waited what seemed to be an hour for him to show up. All at once someone came running to the barn and said that the owner had dropped dead in the house. This was some event and before we could get an undertaker out to the farm to remove the body it was 3:00 o'clock. In spite of it all, we had a successful sale."

Clint Tomson:

"The late W. A. Cochel, editor of the *Weekly Kansas City Star*, felt that registered cattle was the answer and had me work up a group of consignments from Missouri to grade up Arkansas cattle, the animals to be sold in a sale under the guidance of the late Colonel Jack Halsey. Jack's opening remarks were forceful and persuasive, and the audience in casual country attire seated on the banks of a grassy ravine which served as a natural sale ring amphitheater, seemed to be listening intently. This all changed when a white-faced breathless character, waving both arms and gesticulating westward, announced that Harry Bromfield's barn was on fire. Some 170 people, potential purchasers, immediately disappeared from the scene, some by dividing up teams of horses hitched to the fence, others afoot, and many piling into cars which roared away into the bright afternoon. There was a delay of nearly two hours before the customers returned to bid on the cattle, and they didn't save the barn."

Occasionally the excitement of a sale is too much for some. For example, this happened at a Jacksons Mills, West Virginia bull sale. A man bought a bull with Ed Huff, a fine ringman, taking his bid. The fellow immediately got up, walked into the office, wrote out a check for the bull, turned around and dropped dead on the office floor.

This happened at a bull sale in Georgia. A day or so before the sale, the owner had fired his herdsman and installed a new one. The new one, of course, not knowing how the sale bulls had been fed, immediately started them on his own ration which was considerably different to that which the bulls had been accustomed. You've guessed the results–every last bull was scouring–loose as a goose and badly shrunk. "Ham" Hamilton was selling. A big rawboned somewhat vicious bull comes charging in, splattering all over the place at the same time pawing the sawdust like mad. Ham, without hesitation, remarked, "Now let's go boys, here's one that's part cat!"

Bob Taft, a Hereford breeder from Laverne, Oklahoma, is a very funny guy. Bob hasn't had much formal schooling, consequently, would hardly be placed in the same group as English professors. He holds an annual sale of range bulls with most of his remarks during the sale being unusually funny. Here are a few examples:

Hereford men are rather choosy and strict about the color markings and here in America most of them strongly object to any white along the top line which is back of the typically white mane. Occasionally the unscrupulous will pull out any white hairs, neglecting to point this out to unsuspecting customers. Bob Taft is an honest guy and would never resort to such misrepresentation. One year, Bob's best bull had a considerable patch of white hairs right over the loin area. Knowing the objection to such marking, he decided to put this particular bull about last in the sale. However, he being one of Bob's best bulls, a number of buyers ignored the white and as a result the bidding was lively with the bull bringing the top price of the sale. As soon as he left the ring, Bob took the mike and said in his usual droll manner, "If I'd a known you people didn't mind white in the back I wouldn't have bothered to pull it out of the others."

Another time with a very sway-backed bull, and not especially good otherwise, the auctioneer was having difficulty obtaining any bid at all. So Bob's remark was, "This here bull had a real straight back when he was a calf, but I put such a heavy mortgage on him that it pulled his back down that-a-way."

Once in his opening sale remarks, he announced rather apologetically that he felt—"that it is a good set of bulls for the most part, even though there are the usual 'plainer' ones in the lot."

Toward the middle of the sale the auctioneer was getting no action on an animal, so he called on Bob for a speech thinking some little comment from him would help start the bidding. So he handed the mike to Bob—saying, "tell us something about this good bull". "Well," said Bob, "this here's one of the plainer ones I was telling you about."

Although not intended to do so, this honest remark amused the crowd and started the bidding, actually serving as a spark to make lively sales for the remainder of the auction.

An amusing fellow at an auction is the nervous one, especially if it's his first sale. "Ham" Hamilton tells about such a man at a Louisiana sale:

"I was selling near the end of the sale and a most nervous little man was trying to buy a heifer but was limited in the amount he could spend. He became more nervous each time he would fail to buy. Finally his money held out and he was the successful bidder. He jumped up from his seat and began making his way through the huge crowd to the entrance where the cattle were being led out. It happened that the consignor of the heifer our little nervous man had bought was also the consignor of the next heifer in the sale order, so one of his men was holding this next one just outside the entrance. The owner changed animals and headed back to the sale ring with the unsold one. We were holding up the sale briefly awaiting his return.

Our nervous buyer was anxious to get hold of his heifer and was still making his way to the entrance when he saw the man returning to the ring with a heifer. He grabbed the halter and shouted, "You can't sell this heifer, she's mine."

The consignor, not knowing why the fellow was claiming the heifer, figured he was some kind of a nut, so he jerked the rope away. Then the event turned into a wild wrestle for the heifer. We finally recognized what had happened and to the delight of the crowd, I used the PA system to explain to our nervous customer that his heifer had another number stamped on her and was safely tied back in her stall.

When a new breeder consigns to his first "big" sale nine out of ten are usually pretty jumpy. Such was one, a manager for a big Texas Hereford outfit, on his first trip to the Denver Sale.

This fellow led into the ring a bred heifer and like most consignors figured he ought to make a speech to spur on the bidding. So he rapped on the auction box and asked the auctioneer to be heard about "this top heifer."

"Colonel," he stuttered, "this heifer has been examined and is guaranteed to be safe in calf by a veterinarian."

When I was Secretary of the American Hereford Association some wag wrote the following, for which I got credit:

TO: THE STAFF

SUBJECT: DEATH OF A.H.A. EMPLOYEES

It has been brought to our attention that many employees are dying and refusing to fall over after they are dead. THIS MUST BE STOPPED!

On or after the first of September, any employee found sitting up after they have died will be dropped from the payroll at once. (ie. within 2 weeks) Where it can be proved that the employee is being supported by a typewriter or other property marked "AHA" an additional 2 weeks will be granted. The following procedure will be strictly followed:

If after several hours it is noted that the worker has not moved or changed position, the supervisor will investigate.

Because of the highly sensitive nature of AHA employees and the close resemblance between death and their natural working attitude, the investigation is to be made as quietly as possible so as not to disturb the employees if they are only sleeping. If some doubt exists as to the true condition of the employee, extending the pay check is a fine test. If the employee does not reach for it immediately, it may be reasonably assumed that he is dead. In some cases the instinct is so strongly developed, however, that a spasmodic clutch or a reflex action may be encountered. DON'T LET THIS FOOL YOU!

In all cases, a sworn statement by the dead person must be filled out on a special form, 4403A. Fifteen copies will be made. Three copies are to be forwarded to Paul Swaffar's office and three copies are to be given to the deceased. Destroy the rest.

A 22-w application for permanent leave must be filled out by the employee. Be sure to include correct forwarding address. If he cannot write, his signature must be witnessed by two other employees, preferably alive. Complete case by pushing body to the side to make room for the next incumbent.

BY THE DIRECTION OF THE SECRETARY OF AHA

I hereby publicly deny that I wrote same.

Hop Dickenson tells about "Ham" Hamilton and Frazier Biggs at a sale back East. At most farms where sales are being held there are outdoor "Chick Sales", one for Him and one for Her. Usually the door is fastened on the outside with a wooden button while on the inside there is sometimes a rope or wire for the occupant to hold on to or hang on a nail to keep from being disturbed.

On this particular sale day just before time to start selling, Ham and Frazier each decided to "go." Ham went first. When he got to the place, someone was inside holding the door shut with this inside rope. Ham tried the door; but finding the place occupied went around behind out of sight. On the way back to the sale tent, he met Frazier hurrying along.

"Frazier," he said, "it's a nice clean privvy, but the door's stuck so you have to pull extra hard to get in."

Frazier believed the story and Ham hid behind a tree to watch results—he had also quickly gathered others to observe.

Frazier grabbed the door without any warning to the occupant and gave it a mighty yank. The door flew open, of course, with the other fellow hanging on the rope and mad as a hornet. Had he been fully covered he would probably have murdered Mr. Biggs.

At a Shorthorn sale, a ringman while giving a sales talk about a certain bull, attempting to point out that this bull would sire really good calves to feed out. His description went like this: "Boy, this bull would really do a job on a bunch of steers." Clint Tomson without hesitation remarked, "Maybe so, but I suspect he'd do much better on a bunch of heifers!"

Now days, no purebred auction is held without a public address system, usually the louder the better with most every remark made in the auction box being heard by all. Sometimes those in the box forget to turn off the sound or cover the mike when they make private remarks. An example is like the time in a Joe Purdy sale at Butler, Missouri, when an extra large growthy bull came charging into the sale ring. The auctioneer, whom I'll not name, exclaimed rather loudly, "Boy, here's a big bastard, ain't he!"

Another one—The owner was one of those ill-tempered, grouchy fellows who thought every beast on his place was vastly superior to those belonging to anyone else. On this sale morning he had rounded up the auctioneers and all fieldmen, giving them an abusive "hard sell" pressure talk. After the sale had progressed a short time, the auctioneer and two or three others were visiting in the auction box between lots. Over the mike, clear as a bell, came this: "As far as I'm concerned, the old S.O.B. can eat the rest of them!" I've often thought the auctioneer left the mike on purposely.

John Olvey, an old-time Hereford breeder, had two herds—one at Tallulah, Louisiana, the other at Harrison, Arkansas. Several years ago, John dispersed the Louisiana herd with the late Bill Pace as one of the auctioneers. First day of the sale, it rained all day long, in fact, so much that the sale had to be stopped about mid-afternoon. Joe Gossard tells about what happened next morning. "Bill Pace drove in the gate of the parking area and John was standing there directing traffic. Bill's car splashed mud all over John. When Bill noticed what he'd done, he stepped on the gas to get away so John wouldn't recognize who was driving. In doing so, he threw more mud on John. He was a mess. Shortly thereafter, Bill came up to John asking him what had happened. John related the incident. Bill asked who did it and John replied he didn't know him, but he'd damn sure recognize him if he ever saw him again!"

There have been occasions when "deals" were made to make a sale look good on paper. Usually such deals fool nobody and more times than not will backfire on the participants. Here's an example.

A dispersion sale was being held in which the main herd-bull was thought to be an extra good one by the owner and possibly several prospective buyers. The owners felt, however, that they'd like to "tie down" a good price. So did one buyer. Consequently, a deal was made. The bull was sold at $5,000 prior to the sale's start; however, it was decided to run him through the sale ring making him "bring" more money than another bull belonging to a rival breeder had sold for the previous week.

Bidding was lively with several interested breeders actually submitting bids above $30,000. Owners, of course, shed real tears in seeing this $30,000 or more slip away.

Nine times out of ten, this funny stuff catches up. Owners realize it, so just about all has been eliminated in purebred auctions—by the best people, anyhow.

Ed Huff tells a story about sale consultants. Nobody knows exactly what a sale consultant is, but he's been described as a fellow to whom you turn when you don't know the answer—and he doesn't know either.

Here's one about cats. Seems that on an isolated ranch, the owner kept a goodly number of cats in order to help control the varmint population. All except one old Tom were sow cats. This Tom had ruled over his harem for goodness knows how many years having soundly thrashed and chased away each young male challenger. Finally old Tom reached the age and physical condition that he could no longer tend to his duties. The she cats held a secret meeting and decided to obtain a new Tom. However, it was also decided that old Tom just had to be honored in some manner for his loyalty. So, upon motion by the oldest she cat and unanimously passed, old Tom was made a consultant.

Anyway, back to Ed's story. Two rival sale consultants were in the auction box at this particular sale, each doing his best to impress all with his knowledge, salesmanship and promotion ability.

"Here's what happened," relates Ed. "The auctioneer had a firm bid on a heifer at $1,500 when this one consultant allowed it was time to make a speech. He grabbed the mike and with arms flailing the air told the crowd that, 'If this heifer was fed a few bushels of corn she'd easily bring $500.' Naturally everybody in the sale seats howled and the consultant kept quiet for the balance of the day."

Several years back when Johnny Jones worked for the Hereford Association, he and I agreed to help at a State Consignment Sale since no one else could, or possibly would go. The consignment was reasonably good with most of the cattle being properly conditioned and groomed for the sale. However, one breeder had a pen full of bulls running semi-loose, but with each having a rope halter made from ¾″ well rope and dragging some 12-15 feet of rope for leading. It appeared the owner dared not get closer to the bulls. They had been neither washed, brushed nor fed very much, and each tail switch was matted with cockleburrs. Before the sale start when John and I were inspecting the animals the owner's remark to us was: "I never knowed they wanted these bulls all fancied up. If I'd a knowed I'd a fastened mine up."

At this sale, it was a cold miserable day and the crowd was very small. There was very little "ringman help" around, so the auctioneer, Gene Watson, asked Johnny to help out by working one side of the ring. He agreed, working his side earnestly and noisely, but with little success since no one on that side appeared to just have to have a bull.

Right smack in front of John, however, sat two rather elderly gentlemen who acted extremely interested. As each bull came into the ring they would check their catalogues carefully, nod and talk to each other in low tones. Occasionally one or both would get up, walk down to ringside and carefully scrutinize the bull from end to end. Early in the sale John felt sure he had two real hot prospects and, of course, would work hard to get a bid from them. As the sale progressed without success, John became less and less sure.

Finally, toward the end of the sale, Johnny climbed up to where they were sitting; "You fellows better hurry up and buy your bulls, we'll soon be out."

"Why, we don't want no bulls," one said, "we're retired."

One of the oldest and yet funniest gags at cattle auctions is that of "selling" an animal to some unsuspecting person, especially if it's a woman and rather well known.

Ringmen, always alert to the humorous value of this gag, will notice such a prospect—especially if she's talking and waving her hands and arms about. The auctioneer is quickly tipped off so before he actually sells to some real bidder, he'll cry "sold" to the cooperating ringman who will then loudly shout, "Here's the buyer, Colonel, Mrs. Penelope Smalts from Peoria." Of course, Penelope will scream, "No! No! No!," while the crowd roars. Usually then, Penelope keeps completely silent and sits on her hands during the remainder of the sale.

George Morse relates a story concerning Bill Lacy when he worked for the *Hereford Journal.* This happened at a Howard Grover sale out in Kansas.

"Howard had a rather strange sale barn made in an old shed and the ring was of woven wire that must have been at least six and a half feet tall. There was very little room for the ringmen to work outside the ring, so at this particular sale about two of them would work inside. Howard had some bulls that were anything but gentle and it happened to fall Bill Lacy's lot to work inside the ring. Bill was always of a rather jumpy nature and wasn't too sure about competing with the bulls for ring space—nevertheless, he took the assignment. As the sale progressed he became more and more confident. Bill was dressed in proper western attire with a large belt buckle and since it was a warm day he'd removed his coat.

"Gene Watson was doing the sale and had just completed selling a bull which had left the ring. Bill was looking at the crowd with his back to the incoming chute. Howard Grover had warned everyone that the next bull was a little wild, but, Bill, being preoccupied, did not hear the announcement. The bull hit the ring, saw Bill, slid on all four feet and just as he stopped—about three feet from Bill, snorted like a wild lion. Without even looking Bill took a leap just in front of where he was standing and landed with his middle on the wire fence, the big belt buckle firmly engaged in the wire, his head dangling in the crowd and his wildly kicking feet still in the sale ring.

"The crowd roared. It could have been a tragedy, however, the confusion awakened Darrell Sutor who was having his afternoon nap and, therefore, was the only one not doubled up with laughter and who could give Bill some assistance in getting untangled. Bill worked the remainder of the sale on the outside of the wire fence."

OK

Around 1950 Jack Turner, then Secretary of the American Hereford Association, and his Board initiated a policy of hiring newly graduated college animal husbandry majors in a fieldman's training program. I was given the job of screening some of the prospects. Among those hired were George Morse, Max Cox, Gene Brinkman, Bob Felton, Larry Seaman, Kent Mackey and Mag Greenthaller. None are still employed by the Association, but all have done exceptionally well for themselves and for the Hereford industry. The policy then was to start them at rather low salaries and advance them as they progressed.

I'll never forget George Morse coming in for an interview. He had just finished at the University of Missouri, where he had made a good record in student affairs.

Handsome George came in freshly scrubbed, a newly pressed suit, hair neatly combed and after deciding we'd try to hire him, I asked, "How much money do you want?"

"Well," stuttered George, "Pay me just whatever you think I'm worth."

"Oh, Hell," I said, "You'd starve to death on that amount. Maybe we'd better pay you a little more."

Claude Beeks of Baldwin, Kansas, a retired Hereford breeder who, when he was in the business, always exhibited at a number of local shows and on a number of occasions at the American Royal, National Western in Denver and the Kansas State Fair. One year, he had a small exhibit at the town where the Kansas State Mental Institution is located. Each year groups from the Institution were brought to the fair by bus and guided through the exhibits. Claude was seated on a bale of straw behind his cattle when a group came by. One fellow howdied Claude and the conversation went like this:

"Are you a farmer, mister?"

"Yes, I'm a farmer."

"These your cattle?"

"Yes."

"You like farming?"

"Yes, I like farming."

"You ever been crazy?"

"Well, I don't think so."

"Beats farming."

Mike Friend is a good Hereford breeder down in Oklahoma. Mike tends several hundred registered cows by himself, growing out the young heifers and sale bulls with help only at cattle working or haying time. He feeds hay in the winter out of a pick-up truck. The truck is equipped with one of those gadgets that hooks on the steering wheel to keep the truck moving in a straight line in low gear while Mike scatters hay out of the back end. He has an old white dog that always goes along in the cab. The dog moves over behind the steering wheel when Mike gets out of the cab.

One morning Mike was thusly feeding along the side of a country road, with the pick-up moving slowly and the big dog sitting behind the steering wheel. Two local drunks who had been following their trade came by in an old beat-up car, looked over and saw what was going on. Mike heard one say to the other, "Look! That goddam dog's drivin' that pick-up!!!"

JD2

Here's a true story about a buyer who went to look at a bunch of Texas sale heifers at the end of a three-year drouth.

"They're kinda little ain't they?" he asked.

"Yeah!" says the owner, "they're littler than they were when they were little."

O. R. (Pete) Peterson has been one of the characters in the cattle business for some forty years. A Kansan by birth, and a Texan by adoption, Pete probably would rate as one of the most sale-wise people ever to be connected with the Auction business. Although most of his experiences have been with Hereford cattle, he could have been a success with any breed in any species.

Here are a few samples of Pete's stories exactly as he tells them:

Back in the 1930's Harry Warrick and Carl King had ranches adjoining, near Siloam Springs, Arkansas. They were a great pair, generally traveling together. This time, in a consignment sale at Hereford, Texas, there was a bull named Good Mixer which Harry wanted badly, but didnt want anybody to know he was bidding. We got together before the sale and decided that I was to bid for him so long as he had a lighted cigarette in his mouth.

When Good Mixer came in the ring, Harry was so excited he forgot to light the cigarette he'd been carefully holding in his mouth. Of course I didn't know whether to bid for him or not and to chase up and ask him would have defeated his wishes. Carl noticed I wasn't bidding, so he poked Harry in the ribs, handed him a match. Harry was shaking like an aspen leaf, but finally managed to light up and we got the bull bought at around $3,000 as I remember it. Well, Harry was still excited and took off back to Amarillo before the sale was over, forgetting to settle for his bull. When we got back to the hotel we told Harry that the seller, Lee Cochanaur, wouldn't honor the sale because he didn't pay for the bull. However, we told him we'd been able to get another bull for him which we thought would please him better, and that the substitute was being brought up to Amarillo in a pick-up.

In the meantime, Bill McSpadden, a good friend of Harry's, had gone out to the stockyards and purchased a little old runty Jersey calf and slipped it into Harry's bathroom while we had him otherwise occupied. We all went back with Harry to his room and just as we entered this calf let out a beller.

"What the hell's that?" said Harry.

"Oh, it's that silly ass who was stoned last night," we told him.

Harry stomped over, opened the door and here was that little old runt with a sign around his neck, "Good Mixer, Jr."

Harry then realized he'd been "took" so decided to join in the fun, and along with Bill McSpadden, had every bellboy in the hotel come up and open the bathroom door for some trumped up reason. Each would come tearing out when they heard this little Jersey bellow.

Finally they ran out of bellboys so they sent for Mrs. McSpadden's maid who lived in the hotel. She opened the bathroom door, turned white as a sheet, yelled "Lion! Lion!" and then took off in a dead run. They haven't seen the maid since.

Then there was the time at the Denver show when Johnny Dean, Jack Turner and I were rooming together at the Brown Palace Hotel. We all went out to parties on this night, getting back to the hotel rather late. When we opened our door, here were three sheep in the room with signs around each neck reading, John, Jack and Pete. It was years later when we found out that Dan Thornton and a guy by the name of Carey from Wyoming had stolen the champion pen-of-three lambs and sneaked them into our room with some cooperation and an under-the-table-tip to the hotel bellboy.

Speaking of Jack Turner reminds me of the time he, Frank Farley and I were returning a stolen car from Amarillo to Kansas City. This was a *Hereford Journal* car which had been recovered in Amarillo.

On the way we stopped at my brother-in-law's house in Higgins, Texas, so he asked us to return a borrowed gun to another brother-in-law in Kansas City. I told him to load the gun as we might see a coyote while we were driving across the coyote country.

We stopped for the night at Wichita, but since there was a convention we couldn't get our rooms until about 11 P.M. While we were waiting we went out to the stockyards, got a bottle of booze, ate dinner and then went to a friend's house to play poker. When we left, this friend gave us another bottle and asked us to take a girl home who lived across town. This was during prohibition. On the way a cop stopped us. Wichita cops were tough. Here we were

in a stolen car, a loaded gun, two bottles of booze and a girl whom we didn't know, and she was so nervous she couldn't even tell the cop where she lived. Naturally, the cop seeing our nervousness went through our stuff, found the gun and booze, of course, no title to the car.

"This gun loaded?" he asked.

We told him we weren't sure, so he took it out of the car, fired it straight up and about fell over when it went off.

"Listen," he says, "I think I'd better run you guys in--you're in real trouble."

"You're telling me," I say, trying to act nonchalant even though I was scared stiff. "Officer," I say, "We're innocent and I think I can prove it if you'll let me get in touch with some people I know. I used to live here."

"Yeah, now maybe you'd like to tell me who you know."

So I named all the big shots I know. All the while Jack Turner, who was in the back seat, was fumbling through everything he had, including a big check book on a Fort Worth bank, trying to prove his innocence.

Finally, this cop says to me, "Well, I'm going to let you go. I think I remember your face and I believe you're telling the truth. But this Texan, I don't know about him, he looks kinda funny to me."

Jack finally regained his composure, convinced the cop he, too, was honest and we got to bed in the wee hours.

Gene Watson, one of our leading Hereford auctioneers, was at one time a fieldman for the *Hereford Journal*. Like a good many others, he had aspirations to be an auctioneer. Gene was working the ring in the Switzer and Field sale at Gunnison, Colorado, so we decided to let him sell a few tail end lots at the end of the sale. There was one old "shelly" cow left that was too common to even go in the sale. However, we had her brought in and, without Gene's knowledge, of course, "ran" her up and up rapidly to $10,000–she was probably worth about $110. Gene was naturally nervous in his first try at a big sale; however, someone assured him that this cow had a special pedigree and productive history that made several prominent and wealthy breeders want her.

Finally, at $10,000 bid, the buyer's name–Arthur Thompson–the other auctioneer, was turned in. Not till then did Gene know he'd been "took".

Another time, later on, Gene was the auctioneer at a sale early in the fall at Raton, New Mexico. On the sale morning he rode out to the sale barn with a group of us, putting his grip in the trunk of my car as he was to go on to Denver after the sale.

The minute the sale was over we took off for Fort Worth, as there was a storm coming up. We stopped at Dalhart, Texas, for gas and for some reason I opened the trunk.

"Whose grip is this?" I asked. Somebody says, "It's Gene Watson's."

It was Saturday night, so we know Gene is in trouble. We ship the bag to Gene's home, and later find out that he went to a drug store in Denver late that night, bought himself some cheap drawers, a razor, tooth brush and got along moderately until he got his grip the next week.

When I was operating the National Auction Company, one of the auctioneers in my stable was G. H. Shaw. "Gish", as he was popularly known, was one of the fastest auctioneers in the business. For example, one time I bought a herd of cattle down in Texas and shipped them to sell at Statesboro, Georgia. Gish sold 223 head in this sale in 151 minutes.

This particular herd had been involved in a stampede a few days before we shipped them and were naturally very "spooky." Between the time they arrived in Georgia and the sale day we held them on a farm on the edge of town, so we had to drive them a short distance up Highway 80 to the sale barn. About the time they got on the road, a little kid came tearing out toward them, yelling and screaming as loud as he could. Of course, these wild Texas cattle took off in about 18 different directions. Several were killed on the highway, others simply took off to parts unknown and we never did recover them. It was the damndest stampede you ever saw and probably the only one Statesboro, Georgia citizens ever saw.

Oh, yes, I was talking about "Gish" Shaw. One time he and I were rooming together at a big Beeville, Texas sale in November. Sometime during the night I was awakened by Shaw screaming to the top of his voice, "Pete! Pete! Dammit, wake up!"

"What's eatin' on you?" I asked.

"There's a big rat in this room," he says.

We get up and look around and sure enough, we find rat droppings that look like the rat might have been about the size of an alley cat—but we couldn't find the rat. Finally, we discover a hole behind the radiator big enough to admit most any size rat. So we take a big ginger ale bottle, stuff it upside down in the hole and go on back to sleep.

A good many people have had experience with a stray cat crawling up on the motor of a car where it can get warmed up on a chilly night. Well, one time on an Oklahoma tour Mr. W. T. Alexander asked me to ride with him on this brisk morning. There were three or four of us in the car, so since we were all gabbing you couldn't hear anything. Ever now and then though we'd hear the damndest racket—like an animal screaming. Finally the car got hot so we stopped, lifted the hood and here was this dead cat just about chewed to pieces from a bout with the fan belt.

An that reminds me of my good friend, Jim Hering, an old-time Hereford breeder from McGregor, Texas. Jim lives a mile or so from town, but always gets up real early and drives into town for an early "waker-upper" cup of coffee. Jim had left his ventilator window open the night before so when he started up this chilly morning he didn't notice that a big tom cat had crawled in and slept on the back seat. As soon as Jim started the car, the cat took off for the ventilator window which Jim had closed. Jim and the cat went round and round with Jim getting scratched on face, hands, and arms, finally losing control of his car, winding up across the bar ditch and smack into a tree. Cost him considerable to get his car fixed and the tom cat got away without a scratch.

3

There are a number of well-to-do people in the purebred cattle business, and often times you'd never know it by talking to or looking at them. In the summer of 1958, the Fort Worth, Texas Hereford Club held a tour in the area. Late in the afternoon of the tour's first day a good many of those on the tour were lined up for registering at the swank Green Oaks Hotel in suburban Fort Worth. Most of them were dusty, informally dressed and with considerable cow manure on their boots. Two dude salesmen walked past the group and one sneered to the other, "Well, I didn't know they let riff-raff like that in their hotel." The cowmen, of course, ignored the dummies, but later on we figured up that there were seven millionaires in that registering line.

HW

I told about the time Jack Turner, Frank Farley and I got quizzed by the Wichita cop when we were returning the stolen car.

Well, again during prohibition days, when I was the superintendent of the Livestock Show there in Wichita, the boys with the cattle always needed a little drink so we arranged for a bootlegger I knew to walk through the barn and put the orders in the feed boxes. Jim McClelland was showing cattle at the time. The cops got wise and came through the barn searching the boxes. When they found a bottle in Jim's box he put on that innocent look only he can do and explained, "That's not my whiskey, I never saw it before—that ain't even my box."

You know what? The damn scoundrel talked two cops out of running him in.

During the same show, "Lucky" Collins, a reporter from the *Kansas City Star* and a good friend of mine, came in early in the morning after a rough night on the train and wanted a drink. So I took him up on the second floor of the Forum Cafeteria there in Wichita where an undertaker had some caskets stored and in one of which I had hidden a bottle. I opened the casket and gave him a drink. Lucky says, "What am I supposed to do, take a drink of this stuff and then crawl in?"

Then one other night, after the champion bull had been selected, the owner agreed to treat all the boys with the cattle. I arranged for a back room over this same Forum. The room had only one door with all the windows barred. While the boys were in there, I arranged for two cops, friends of mine, to raid the place. One was a big cop named Pete and the other a little guy. The big one walked in, leaving the other one at the door, and announced that, "This is a whiskey raid." Les Mathers, an old-time and popular Shorthorn breeder was in the crowd, and later said to me, "If those windows hadn't been barred, I'd have been wearing a frame around by neck yet, because I'd have gone right through one."

After a few minutes I couldn't hold it any longer and had to laugh. The cops joined in and I had quite a little trouble talking the gang out of murdering me.

When I was a student at Kansas State, one of my professors was A. M. (Andy) Patterson. Andy was a big tough, ugly, lovable guy who later became manager of the American Royal in Kansas City. He was and still is one of my best friends. Andy was a great practical joker, especially if the joke involved me or my wife, Mary. When I was superintendent of the Wichita Show, Mary and I always went in to Kansas City to attend the American Royal. This particular year Andy and his wife, Janet, invited us to come and stay with them.

The first night there we sat around and talked quite awhile, so Mary decided she was tired and would retire before the rest of us. After she'd gone to her room, Andy put on his short nightgown, then barged in and jumped right in bed with her. She let out a scream you could hear for a quarter of a mile, and Janet and I witnessed the damndest hen-rooster race that Kansas City ever had.

One year we sold a bull for Dr. Charles Harris in the Fort Worth sale for about $8,000 to an Oklahoman whose name I have forgotten. The next year in this sale we were having trouble getting $2,000 on another of the Doctor's bulls. My smart aleck remark was that the year before we had sold a Harris bull for $8,000 which was not half as good as the one in the ring. Unbeknown to me the purchaser of the $8,000 bull was in the stands and he had scheduled a sale featuring his bull. This fellow came charging down in the ring and dared me to come outside to settle this thing. I had one helluva time convincing him my remark was supposed to be a joke.

Another time, when Dan Thornton decided to disperse his herd at Gunnison, Colorado, he was kind enough to ask me to help him. Mary and I went up there and spent several days lining up the arrangements. We stayed at the hotel which was managed by Ted McCorkle. We went back to Fort Worth, where I got the catalog finished, placed the advertising and then went back to Gunnison without Mary.

In a day or so Mary drove my car up, bringing along Harry Holtz, who was to clerk the sale, and Harry's wife. When they arrived at the hotel Ted McCorkle, who, of course, knew Mary very well, was on the sidewalk in front of the hotel.

"Well, hello, Ted," says Mary.

"Hello, but who are you?" asks Ted.

"Why I'm Pete's wife."

"Pete who?"

"Oh, you know, Pete Peterson," says Mary.

"You can't be Pete's wife, she's been here every night for two weeks."

Mary was a bit shocked, but she got the key and went on up to the room in which I was registered. There were women's dresses, shoes and stockings around the room, and women's unmentionables hanging in the bathroom.

Dan Thornton and McCorkle had really given her a warm greeting.

I have always liked the auction sale business, and there are no finer people on earth than the cattlemen. My father took me to my first auction when I was about twelve years old and I've been connected in one way or another with the business for over forty years.

During that time I've known most of the good auctioneers in the country and I've always had great respect for the unique ability these fellows have. I'd rate Art Thompson as the greatest I've known. From there on we've seen so many good ones I could not possibly rate them.

Since 1946 I've had my own company, have dealt with many, many sales and have managed a number where the proceeds totaled up to a million dollars. During that time I've had only two written contracts. When a cattleman gave you his word no other contract was needed. I do not know of any other business anywhere in which so much trust is placed in a man's oral agreement. The business has been extremely kind to me, and I've handled several sales where my fee ran well into five figures. The highest fee I ever received was from the LL Jones dispersal at Garden City, Kansas. However, in this one I paid for all advertising, cataloging, auctioneering and most incidentals. My highest net fee was from the Honey Creek dispersal in Oklahoma in which the cattle sold for over a million dollars.

CHAPTER 13

GOOD SENSE—OR THEN, NOW AND WHAT'S AHEAD

Chapter 13

Good Sense—or Then, Now and What's Ahead

The material in this chapter was contributed by four of my long time and admired friends. Each, however, is more than a personal friend. Each in his own way is an able and dedicated friend of the livestock industry. These men are experienced, knowledgeable and sincere people who have contributed much toward improving animal agriculture. This chapter contains a wealth of good common sense and I am grateful and proud to have their contributions. It is doubtful that there are any four men who have more friends and admirers in the cattle industry.

Don Ornduff

No person in all of America knows more about Hereford history —people and cattle responsible for building a breed—than does Don Ornduff. His painstaking collection of facts amassed in the fine book THE HEREFORD IN AMERICA is a masterpiece of breed history and belongs in the home of every Hereford breeder and on every agricultural library shelf in America.

Don Ornduff grew up on an Iowa farm where he gained his love for good livestock, especially Herefords. He joined the staff of the American Hereford Journal in 1930, becoming editor of that publication in 1944, a job he held with great distinction until his retirement in 1970. Perhaps no man in all the world has written more about any breed than has Mr. Ornduff written about Herefords. In all that time, every word he wrote was constructive, truthful and sincere—no vicious, gossipy or "far out" slanderous, sensational journalism so common in some places and which most of us abhor.

Since his retirement, Don has been busy as usual, writing, working and contributing to a breed and an industry he has always loved.

DONALD ORNDUFF

The most visible changes relating to beef cattle during the years of my observation have been those associated with size. Trends which have followed each other during the period dating back to the 1920s produced a roller-coaster effect, with weights ranging from up to down and back to up. There have been exceptions, of course, not only in individuals but also in herds and families, but, in general, this has been the rule.

Nor is this something applicable only to the middle years of the 20th century. It is an historic fact dating back to the Hereford breed's beginning years in the middle to latter 1700s. The native cattle in a remote part of southwestern England which provided the Hereford foundation stock were big cattle. They served as draft animals in tilling the soil, not unlike some of the latter-day kinds from Europe which continued to be used for such purposes there until recent times. They were big because they had to be big to pull the plow. And then after a half-dozen or more years at the yoke, they were sent to slaughter to provide food which even then was considered better than that from any other source.

With this evidence of the functional virtue of size, it is not surprising to read in the agricultural histories that the English farmer-stockmen emphasized size in the early development of the breed. Weights of 3,000 pounds and more were considered not at all uncommon for Herefords. For example, Old Wellington, calved at the beginning of the 19th century and sired by Silver, a key sire in the breed's early development, weighed 3,012 pounds. The William Hewer herd, in which Silver was bred, had another bull, Hamlet, which was reported to have weighed 2,800 pounds at the age of 2 years and 6 months, while a Hewer steer had a recorded weight of 2,900 pounds. That this tendency toward size and scale continued well into the 19th century is evidenced by the fact that the English bull, Cotmore, which was calved in 1836, is listed as having weighed at maturity 3,920 pounds. The English Hereford historians, MacDonald and Sinclair, in 1909 recorded that Cotmore probably was "the heaviest beast ever bred in England," and added that he was "as good as he was great." This appraisal was buttressed by the fact that Cotmore won first as a 2-year-old, as a 3-year-old and as an aged bull in the cattle show at the city of Hereford, and that he also won first at the English Royal in 1839.

It is of interest, and perhaps significant in indicating the role of man in planning the matings that lead to developments toward specific goals, to note that Cotmore's grandsire was a grandson of Old Wellington, the 3,012-pound bull noted above. This is indicative of the fact that man's guidance in the planning of matings and, thus of inheritance, is a determining factor in the direction of a breed's development and further denotes the historic responsiveness of the Herefords to man's direction. The widely recognized element of adaptability or versatility, which has endeared the Hereford to successive generations of stockmen, always has enabled the breed to respond readily in such directions as the owners set their minds to.

But not always have men directed the breed with intelligence and well reasoned judgment. This has led the Whitefaces on occasion to take wrong turns which have temporarily slowed their progress. Thus, the very fact of their easy adaptability and ready responsiveness to breeders' guidance has at times reacted against the breed's own best interests. That remarkable progress has been recorded, however, during the past 200 years of Hereford history is sufficient evidence that breeders, in general, have been right more often than wrong.

So it is apparent that size is no discovery of the middle years of the 20th century, despite exhortations in behalf of some of the behemoths of various colors that made their appearance on the bovine stage during this era. The big Herefords of the breed's beginning years were gradually adjusted downward in size as stockmen sought to strengthen the breed's beef-making qualities. In return, there came a gain in smoothness, heavier shoulders and hindquarters, a thickening of the covering of flesh and a refinement of bone. Because the breed's primary function no longer was pulling the plow.

With this change in emphasis, as devised and reasoned out in the minds of men eager to supply the demand for food which rose with the industrialization of England, came greater facility for fattening and early maturity. The Herefords were well along on this course by the time the heavy importations came from the breed's homeland to establish the Whitefaces as a major beefmaking force on this side of the Atlantic.

For a while the Americans were content to follow the pattern which had carried the breed to its English successes. But restless minds continued to probe for possibilities of change, more or less toward added refinement which sometimes proved to be at the expense of size. Unless it is at least periodically reinforced, size has a way of deteriorating, especially in times when there is a tendency toward making cattle smoother and perhaps more pleasing to the eye—"prettier," in a word. So, almost imperceptibly, there was some decline in average size, but with some exceptions, of course.

Anyone who has watched the swing of a clock's pendulum knows that it goes from extreme to extreme. It has been much the same with the matter of size in beef animals. By the 1920s there were some who were sufficiently concerned with what they deemed to be too much of a drift toward a smaller-size beast who sought to rectify the matter by accenting sharply increased weights in Herefords. One herd developed and publicized what were termed "big-type" Herefords. Herd bulls publicized over a period of years ranged from 2,600 to 2,865 pounds, and again the influence of man's mind and guidance as reflected in selection and matings was evidenced by the fact that the bloodlines of this herd differed but little in origin from those of most herds of its time.

Another similar example was offered in the instance of a bull which came out of Kentucky to make a "big" impression on the crowds at some of the shows of the latter 1920s. He weighed 2,800 pounds, and what to do with him in a show class was something of a puzzle to judges accustomed to viewing bulls of his age scaling markedly less. He, too, came from bloodlines not greatly unlike those from which sprang showring competitors weighing so much less that they might have seemed to a novice to be of another breed. So, once more, it may not be inappropriate to note that the overly large are not an "invention" of recent times. They are merely another manifestation of the extreme upward swing of the pendulum.

What interest there may have been at that time in turning the breed as a whole in the direction of more size was arrested by another development which came to the fore almost simultaneously with the appearance of the 2,800-pound show bull. It was the concept of baby beef—that is, beef from bovines of smaller size than those from which beef had mainly come in the past. It was emphasized, both in official and unofficial circles, that cattle of such optimum size would mature more rapidly and yield smaller finished carcasses than had been in vogue previously. It was explained that this was in line with the desires of modern housewives who had declining interest in the large cuts which were in demand when families were larger and the man of the house was more likely to be engaged in pursuits involving manual labor.

Thus was sparked new impetus in the direction of smaller mature size in beef cattle. The word, "tidy," came into the cattleman's vocabulary as descriptive of the size and type of animal which came into substantial popularity during this period. Advocates of this movement also pointed out that the cost of maintaining a smaller animal was less than for a big one, and then there was the matter of eye-appeal. Americans became aware as never before of appearance, and advertising promotion, public relations campaigns, the new picture magazines and the like seemed to make the point in all walks of life that what looked good was good. And more than a few cattle operators seemed to accept that verdict.

Again,. it was the mind of man, or a composite of the minds of a substantial number of men, that guided this change in the Hereford make-up. The breed merely responded in implementing man's decisions. So it may be only fair to state that such problems of this nature as the breed has experienced have been mainly attributable not to the cattle themselves but to decisions to follow whims or fancies of no more than passing vogue.

The ultimate folly was recorded in the early 1940s with the advent of cattle that were so extremely compact in appearance that the word, "compress," (meaning compressed) was coined to identify them.

This type was not at all unique to the Hereford breed, as other kinds also had their representatives, all hailed by their backers as likely to lead the way to "the promised land in the beef cattle business."

These cattle were so blocky and short-legged—giving them an over-all height of from six to eight inches less than others of the same age—that they were, as one writer put it, "almost as different as another breed of cattle." Described as "dumpy" in appearance, it was generally agreed, even by their advocates, that they were small, but to compensate for this it was said that these "compact cattle would carry more beef in less space."

This was a theory which most practical cattlemen and industry observers refused to accept but during this entire period little had been accomplished toward reversing the general tendency that had set in with the advent of the baby beef era. And then an unrelated genetic phenomenon known as dwarfism struck the industry—Herefords and other breeds alike, and involving by no means only parent cattle of the smaller size—with disastrous results. Drastic measures during the ensuing decade virtually eliminated this affliction as a factor of consequence in the beef-making industry, but the trauma of that experience definitely was a factor in again turning the industry in the direction of decidedly increased size and scale. The resultant economic advantages, by then being documented, were being trumpeted, especially by the big-scale feedlot operators, whose voices were being increasingly heeded. So impetus was given to an upward swing which had not yet reached its zenith as the decade of the 1970s got under way. Once again herd bulls weighing from 2,300 to 2,600 pounds were being so extensively publicized as to no longer be regarded as of uncommon size.

Even if repetitious, it may not be out of order to point out once again that the cattle at both ends of the weight scale sprang from the same foundation stock. In other words, the breed responded to the selection practices of men who were seeking to produce larger cattle—or smaller cattle—and who chose the breeding stock and concentrated the genes deemed most likely to yield the desired result.

This is nowhere more apparent than in the "big" bulls from Canada that have come to the United States since the 1950s. A preponderance of them are descendants of U. S. bloodlines that went to Canadian breeders in the 1920s and '30s. Those Canadian breeders and their successors, with an eye to the size and ruggedness regarded as a necessity in standing up to the harsh conditions in that far-northern country, mainly selected through the years for those very factors which American producers actively sought again as the turn to heavier weight came. Thus, from the very same original sources came the cattle which mainly prevailed at both ends of the weight scale, depending solely upon the selection guidance of the breeders involved. All that the breed did was respond to the owners who directed its destiny. And, indeed, it might almost be said that on occasion they did this too well.

No one with an awareness of the cattlemen's independence of mind ever would expect general agreement among them on the matter of size. Most of them probably would agree, however, upon the desirability of combining in their output the gaining ability of large cattle and the finishing ability of smaller cattle without sacrifice of efficiency of gain, if that were possible. But then there are variables within that concept, even if it were fully within reach.

A great philosopher once said, "Moderation is best, and avoid all extremes." Advocates of medium size in beef cattle obviously are of that persuasion. They also have a practical point of view in explaining their convictions in this direction. By emphasizing medium size in their output, they say, they have economic opportunities not available to herd owners whose cattle are of sizes regarded as either very large or too small. These come about, the owners add, through the fact that they have cattle ready to sell on both swings of the pendulum—either when the general weight trend is up, or down. Their cattle, they assert, not being at either extreme, are never very far from the optimum. Which is fairly well in line with Nature itself which holds pretty well to the center of the road. Nature, by and large, is reasonably conservative, and would seem most likely to cooperate best with the cowman who seeks to "get 'em big enough but not too big."

In looking to the future, it might be well to bear in mind the words of Winston Churchill, who once remarked that "The farther back you can look, the farther forward you are likely to see." Or the comment of George Santayana, a famous writer and philosopher, who once warned: "Those who do not remember the past are condemned to repeat it." Animal scientists will give unprecedented guidance to the beef-cattle industry in coming years, and it is to be hoped that there will be among them an awareness of the route by which it has arrived at this point in history. If there is, this should lend to the practical application of future scientific developments of usefulness and value to beef-cattle production.

No attempt will be made here to detail current research and experimentation applicable to the beef cattle field, nor even to recount what has been accomplished in recent times. It is far too voluminous. But it may be significant to note that some 80 percent of all of the scientists in the history of the world are living yet today. The percentage among strictly animal scientists may well be even higher. Bear in mind how recently the change was made from "Department of Animal Husbandry" to "Department of Animal Science" in most of this country's educational institutions which are active in this field.

Cattlemen of the future will capitalize upon this scientific potential. In concert with past experience, whether their own or that of others which is applicable, this will provide new opportunities upon which future cattlemen can capitalize.

Already much in use is progeny and performance testing which, like all agricultural innovations, is a result of scientific evolution, in this case implemented by computer techniques. The development of this and related programs, such as carcass performance, may tend to stabilize the factors of optimum size and desirable type in beef cattle as they have never been. Research obviously will continue and facts not yet known will be filled in to the extent that even the reluctant individualist probably cannot long afford to hold out. Economic necessity will dictate his participation if he chooses to continue in the beef-cattle business.

A goal worthy of all-out effort is consistency, the production of good cattle in quantity, rather than the further improvement of today's best cattle. If this can be accomplished, the matter of size automatically will take care of itself. It will be the size which will be most consistently profitable at every step in the process of beef production and marketing, right down through retail channels. Possibilities along this line are evident in a 20-percent increase in feed conversion efficiency since around 1960 when commercial cattle producers and feeders dictated the pendulum swing back toward larger cattle.

There are other elements, some only now developing, that will further tend to make the cattle business of the future a different "ball game." Specification buying of beef may be extensively developed in the years not too far ahead. Some cattlemen already are looking forward to collecting a premium based on cutability and yield grade of their cattle, in addition to the standard price per hundredweight that the packer pays. Another factor might be the flattening out of peaks and valleys in marketing volume, to avoid the gluts which at times in the past have demoralized feeder cattle and fed-beef prices. If investors from outside the industry increasingly should come to play a role in beef production, cattle financing operations may well take a different turn, with possibly far-reaching effects. There is also the possibility of vertical integration in the beef cattle business by packers and supermarkets, which, if unchecked, could bring economic pressure on both producers and feedlot operators.

This rather sketchy resume merely indicates a few of the possibilities and complexities which may confront cattlemen of the '70s and beyond. That there will be others is beyond doubt, and it is just as certain that future cattlemen must be students in depth not just of their own industry but of all elements relevant to it—or at least to have access to such findings as may be beneficial to them.

While it is evident that the performers-oriented Hereford industry of the future will utilize tools unavailable to previous generations of breeders, it is equally certain that its accomplishments will spring from the heredity base that has evolved during the past hundred years. The foundation for the Hereford breed of tomorrow is that which presently exists. It can never be otherwise. Ignoring the past is like amnesia to the memory. "Without it," as Bishop Fulton J. Sheen has said, "we are likely to think that we have original ideas." The evolution in future generations of Herefords, as already cited, will be those which result from changes in direction in the emphasis that breeders themselves will apply in striving to hasten the progress toward such goals as new procedures and the economics of the beef-making industry of tomorrow may dictate.

But in looking ahead, it is important to continue to bear in mind that neither in the cattle industry nor any other field does the future depend solely upon science and technology. There is always the human element. While man can evolve science and plan its application, can any cattle producer even imagine that his role will end when he pushes a button, or even a whole row of buttons? Even in an age of science, man guides the capsule; man walks on the moon. His spirit, ideals and judgment are still indispensible.

The human element also figured in the personal relationships which have made the Hereford business such a pleasurable activity for generations of breeders. At shows and sales, on farms and ranches, wherever two or more of them got together, interest and enthusiasm usually mounted as their favorite subject came under discussion, as it always promptly did. There was an easy comaraderie as these men became members of a fraternity dedicated to the improvement and advancement of their breed, their industry and its interests.

For them it was both a business and a way of life, such as perhaps it never again can be in exactly the same way. It is now much more in the nature of a strictly "big business" operation guided more and more by "big business" considerations.

Just as today cattlemen are in an era of push-button record-keeping and cattle-feeding, their own travels, except locally and mainly of necessity, are via 600-mile-an-hour jet planes rather than on the trains of yester-year. How times have changed since this writer's own early contacts with the Hereford industry of not so long ago! Sale catalogues in those first years carried railroad time tables for the convenience of those traveling to the sale. Frequently a score, or perhaps many more, would travel as a group between sales which often were set up in a series, a practice which yet is occasionally followed. But the trips now are so quickly accomplished that something of the fellowship, and the yarns that followed, are lost.

One true tale dating back to the old days which comes to mind relates to two then widely prominent Hereford breeders who were in a group attending an auction sale in the Kansas City area. One was a man of considerable dignity, and perhaps vanity, who wore a toupee. Seated beside him on sale day in a chair section adjacent to the ring was his fellow Hereford raiser and personal friend, who had an artificial limb. The sale was moving along routinely until, at length, a cow, apparently panicked by the noise and confusion, surged out of the chute and charged straight toward the section in which the two sat. A mad scramble for safety ensued, especially among those closest to the fence that surrounded the ring. When it was over, with the cow still on the inside, chairs were in disarray in all directions. And, almost lost among them were the two friends who had fallen ignominiously in their attempt to flee, the toupee of the one perched jauntily on a slanting protrusion which proved to be the wooden leg of the other. Except perhaps for wounded dignity, no damage had been done.

The crowd, apprehensive only moments before, chuckled a bit, soon settled down to business again, and carried the sale on to a successful conclusion. Just so, it has always been in the Hereford business. Sometimes problems, sometimes confusion, but always a buckling down in working things out. And then onward again.

Habitually in the Hereford business, as in this case, successful conclusions there have been. In the years ahead there will be new problems, but there also will be new tools and new solutions. This, in the light of the past, lends confidence to the outlook.

B. C. Snidow

I have known Bud Snidow since he was a freshman at Virginia Polytechnic Institute. He was a fine young man at that time and he has grown in stature with each passing year. For six years following his military service, he was editor of the old EASTERN BREEDER *magazine, resigning that position to become Eastern Representative for the American Hereford Association in 1951. Today, Bud is Assistant Secretary of the Association where he capably handles any assignment which comes his way.*

So far as I know, Bud Snidow hasn't an enemy in the whole wide world. He has made thousands of friends for Herefords by his quiet, efficient, friendly and many times gently humorous ways. In the Hereford industry, when someone demands that a job be done RIGHT, *it's "turn it over to Bud".*

B. C. (BUD) SNIDOW

Across my desk has come miscellaneous literature among which is put-together material from several "authorities" written about stock shows.

Stock shows in general are of considerable interest to me since I have been associated with more than just several in the course of my life's work.

I am acquainted with some of these authors, so I was curious about what they were writing. As I read this material, it wasn't long before I got the impression that some authorities are firmly and even bitterly convinced that the show rings are entirely to blame for whatever they think is wrong with beef cattle. This is a big surprise to me for I often have heard that the stock show is the greatest influence for livestock improvement in American history. However, these pseudo authorities claim that shows are misleading and detrimental to real progress in cattle improvement. If this is true, how have we had such good cattle all through the years and how have we gotten along so good using herd bulls that were champions and prize winners at the shows? We all recognize that fads and fancies come and go and that there are mistakes and tricky showmanship used, but I am sorry to learn from these "authorities" that all breeders and exhibitors in the past made no real progress.

Now I recollect that most of the time when I went to a show I bumped into many of these authors and they were there at ringside just like I was, except, of course, when they were out in the middle doing the judging or otherwise expressing their leadership. This raises another question in my mind—if shows have been the cause of what led us to so-called decay, how come these knowing sort of fellows let it happen? If some were around doing the judging or otherwise being big shots, didn't they lead the way? Couldn't they have headed off the situation which has brought into use such phrases as "no economical importance," "need for more length," "greater weight-per-day," et cetera? So it is obvious that with all their learned wisdom these authors might have the human failing of better hindsight than foresight—should their criticisms prove valid.

Everybody in his right mind will agree that some mistakes take place at shows and some of the practices in fitting and showing are silly. Now and then a counterfeit critter will get to be a champion and even on the Register of Merit list. Some "no-goods" bring the highest dollar at auctions and some judging may favor extremely compact "pony" cattle resulting in terms like "show type" and "commercial type."

I can recall some of the cattle of the 1930s and early 1940s. Many of them were rough, hooky, leggy critters that needed changing. They needed some shortening, smoothing and "modernizing" for their time. But conditions were different then with all the professors and market men preaching for early maturity and for an animal that fattened rapidly. The market at that time paid the most for the fattest cattle (and often still does). If some breeders went too far in selecting for these traits and if they demonstrated this accomplishment at the shows, where it was awarded, they perhaps deserve at least a left-handed compliment in acknowledging that they did their job too well. Some went to the "extreme" and "extreme" is what generally causes the most troubles.

Those who criticize the past—some of whom are doing lots of today's judging—are on mighty thin ice for some of the steers they now select. Some of the bulls and heifers they favor now look mighty "extreme" toward the thin side. Practical evidence shows that you sacrifice cutability when you move past body type No. 4 (See illustrations of body types below). Successful cowmen everywhere back away from the slim-waisted, spindly-looking cows and heifers which so-called "progressives" seem to favor these days. They know from long experience that cows without some middle and some ruggedness just can't stand a tough winter and just don't raise good calves.

While we are about the business of condemning shows as being responsible for too much shortness of leg, short bodies, excessive fat and whatever else is wrong, we ought to recall that for years we learned to appreciate such traits, not only at shows but in classrooms or in judging contests. Remember "I put this steer at the top of the class because he is the fattest steer in the class"? There was no better reason to use in a "fat steer class." Generally you scored real well when you correctly said, "I place this bull first because he is the shortest legged, shortest bodied, deepest bull in the class." Reasons such as these were right for their era because cattlemen then were called on to change their cattle to meet the market demands of that day. One day it will probably come about that our contemporary teachers and breeders are in for some criticism, for we now hear lots of kids in contests and lots of breeders, too, using such terms as "the longest steer in the class," "the bull with more length of leg,"

BODY TYPES

Illustrated in outline drawings are the body types found in Herefords, ranging from the compact, lowset Type 1 to the lengthier, taller Type 5. The majority of cattlemen today prefer Type 4.

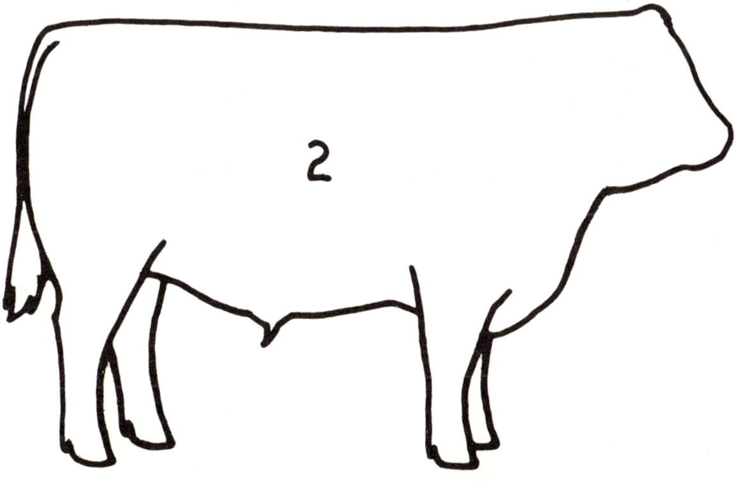

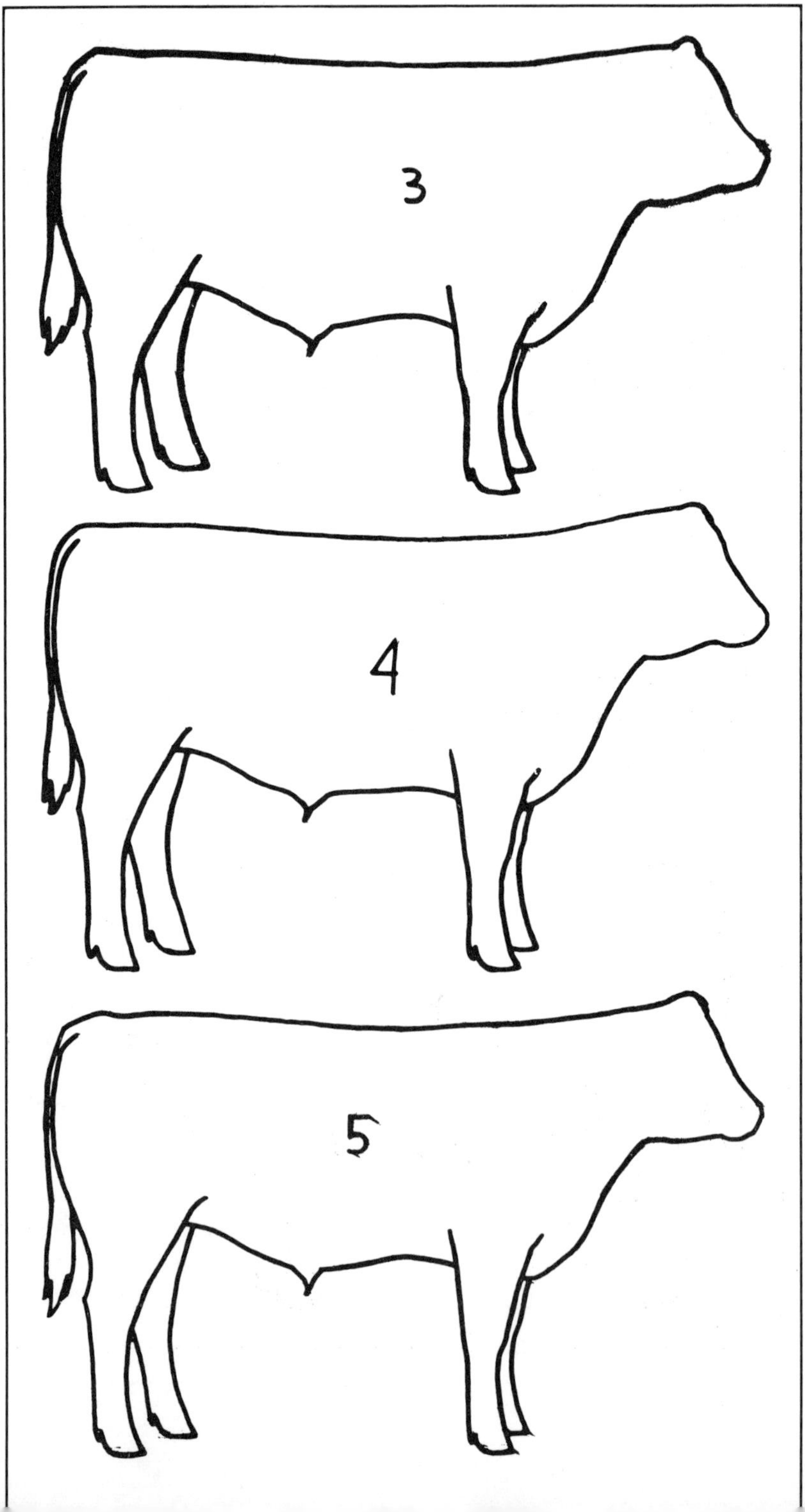
3
4
5

"the greatest size,"—but rarely do you hear an expression of "adequate" or "enough." Sure as anything, we are headed again toward extremes.

It seems to me that instead of harping on past mistakes it is wise to recognize them and to use them as experience and thus avoid reoccurrence in the opposite direction. It is said that history repeats itself. Are we not smart enough to recognize its lessons? Much as I don't like to admit it, it looks as though we are headed toward some of the mistakes of the past although in different directions. Cannot shows be used as instruments in avoiding mistakes, in demonstrating elements of extremes and pointing out practices and possible trends in which we are heading for trouble? Isn't that what shows are for as well as to exhibit accomplishments? If so, then some of the critics of past shows who are doing some current judging should do some soul searching to see what is happening as they help to bring around a transition in beef type. They bring emphasis on certain characteristics that fit in the "modern" concept and while some of this we need, we don't need the "extreme" which some are advocating.

Some of the present critics of shows seem to have created the impression that we were so misled we must now do a complete about face, we should go back to the kind of cattle we had before the 1930s. However, it is true that some of the cattle then would still be good cattle if around today, but to many now, the idea seems to be to revert to cattle that simply are tall and long. The fact that along with this long, tall kind will come slim bellies, thin muscles, cat hams, etc., seems to make no difference because it is called "trim." A snake could be called "trim." Remember when someone praised a bull simply because he was short legged and the counter remark was "even a snake is low down"? I am betting this race horse type won't last long. The feeders complain about them and say they don't want to buy them. That's enough for me.

While we are about it, let's take a long hard look at some of the present day practices that some modern critics of the past seem to tolerate and in some cases encourage. Let's look at the modern foolishness. In their anxiety to achieve trimness of middle some smart showmen of today are accomplished barbers. Next to the electric fans (which are needed) the biggest noise in the barns comes from electric clippers. A herdsman's time is mostly spent in shearing hair off the bell, flanks and brisket. In times past they groomed the hair to accentuate width and depth. Now they shear it off, so maybe they actually accomplish more in the art of barbering than they do in changing the type of animal.

One cattleman from the Dakotas voices his opinion of the practical value of shearing the bellies of show cattle as follows: "If we did that to our cattle they'd freeze to death for sure."

Worse still is the practice among modern showmen of "shrinking." Practical beef production usually calls for feeding for maximum gain. Shrinking doesn't have a place in a practical feedlot, but it seems to be the "in" thing now. Cattle need exercise and I guess we could tolerate jogging, but nowadays we see show steers actually running to keep their weight down. And, instead of giving a steer all he will eat, today's steer exhibitor is often withholding feed and especially just before they are shown, steers are held off feed and water so they are shrunk and dehydrated by show time. To keep this hungry steer from eating anything he can find, including his bedding, some are seen wearing muzzles! Add all this to a slick barbering job and it's no wonder some of our show steers are said to look like "gutted show birds"! These steers look trim all right, but the truth is they are starved. Is it any wonder that so many don't make the Choice grade in the carcass? Some judges who make grand champions out of such specimens are among those who ridicule shows of the past.

As one gets a bit older and can remember his own experiences and mistakes perhaps he becomes more tolerant. I often recall the words of the late Dr. A. E. Darlow, who summed up the situation pretty well when he said, "Maybe stock shows have not accomplished all they might have and all that their managers wanted them to, but then, our preachers and our churches certainly haven't accomplished as much as they would like to do either." The thing is to recognize the good that has been accomplished by the preachers and the stock shows, too.

Maybe we will simply have to say that with stock shows the old slogan applies, "There's no business like show business." We should recognize reasonable styles in grooming and such other practices included in good showmanship. But there is no way that we can justify representing hunger as a substitute for a steer becoming "over done", and for substituting a skillful job with clippers for actually breeding off excess belly and brisket, and for selecting thin muscled, light boned cattle simply to get length and height at the expense of solid substance needed in beef cattle. So, today's leaders should be careful about going to the extreme lest they be subjected to the same criticism that some of them are now leveling at those of the past. Probably we should give more attention to marketing on time, thus avoiding an over done steer or an under done steer.

Criticism can become a habit and apparently those who are the strongest practitioners in panning livestock shows have had some success in pointing out a few of the practices and a few of the cattle that in the past have not been constructive. Apparently impressed with the attention they have drawn, they have kept at it and have expanded their criticism and made their voices louder. They have the habit. Perhaps they should take a closer look at the good things that have happened at shows over the years. For instance, the extreme pony type cattle are gone from the show scene, and so are the excessively fitted cattle, especially females. The two-year-old and senior yearling heifers are not at the Hereford shows any more—they're at home raising calves where they should be. And there hasn't been a nurse cow allowed on the show grounds for years. The good top cattle of today don't waddle. They walk and move out like a stud horse. The show ring of today is a whole new ball game with placings made on a combination of human judgment through visual appraisal, plus factual information of weight and weight-per-day of age. And the audience likes it this way. More and more commercial cattlemen are returning to the ringside to appraise the purebred breeder's accomplishments as the source of good breeding stock.

Without a doubt, and with all due respect to new tools and technology at the disposal of today's cattle breeder, the show ring has been the greatest single force in livestock improvement through the years. There is no reason why it should not continue as a highly useful instrument in education, demonstration and merchandizing in the livestock industry of the future. The mistakes of the past should be lessons not to be repeated. We should accent the positive and the good that shows contribute. After all, past errors were not due to the show ring itself, nor to the cattle, but instead were due to the PEOPLE. The livestock business is a "people" business, so the challenge I make is this; Instead of running down the people of the past, try to build on and expand the good things they accomplished, which we, in all honesty, will have to admit HAVE BEEN RATHER REMARKABLE.

196 *Harold Thieman*

Harold Thieman of Concordia, Missouri, raises, talks, eats and sleeps Polled Shorthorns. Although he sincerely believes that his breed offers more than any other, he is nonetheless a stockman first.

A charming gifted fellow with a twinkle and a quick wit, he is admired by all who know him. He has successfully shown cattle since he was a mere lad and has judged all over America plus a few overseas assignments. I know everyone will be interested in and will enjoy this forthright article.

HAROLD THIEMAN

Shortly after the turn of the century, I remember Dad bought about sixty head of southern cattle. They were three years of age or older and were bought at about seven cents a pound delivered to our stockyards at Aullville, Missouri. They were actually bought at so many dollars per head. During this era every town, large or small, had a local stockyards.

The steers could travel handily, had absolutely no back fat, but did have an abundance of frame, hide, and bone. All of these characteristics were quite evident when given the eye-ball test. Dad, a hired hand, a good neighbor, my brother and myself, rode our horses and ponies and drove this set of steers to our feedlot. They were watered and fed hay and rested for the night.

The cattle were fed corn silage, clover hay, salt, and a mineral mixture. They gained an average of over six pounds per head per day during the next sixty days. When they went to market they sold at fourteen seventy-five per hundred weight and, needless to say, they made us plenty of money.

Before my time, in 1900, a dentist at Lexington, Missouri had a small herd of Polled Shorthorns. He had been losing a few chickens every night for a while so he rigged an alarm on his hen house. The alarm went off so the dentist fired his shotgun twice to scare the thieves. Instead of scaring them, the thieves killed the dentist. They were later caught by a group of civic-minded neighbors and promptly hung from a venerable old roadside tree. Shortly afterward the dentist's widow held an auction, selling everything and that is when the Thieman family bought their foundation cattle in the Polled Shorthorn business.

Shorthorns were the first purebred cattle to be imported into the United States. They and their influence rapidly spread westward and became very popular because they did a tremendous job of improving the quality of beef when crossed on native cattle. They were good-doing cattle of gentle disposition, and gave an abundance of milk. There is little doubt that the early widespread use of Shorthorn blood, set the stage for the use of Hereford bulls in cross breeding. The resultant calves were the best around.

The Polled branch of the breed started from polled "sports" of straight horned ancestry and for some time the Polled cattle were selected mainly for the polled feature. It is only fair to say that many of the early Polled cattle left a lot to be desired in a beef animal. Even so, the Polled feature is dominant and the breed improved rapidly.

In the early 1920s Dad bought a Polled Shorthorn bull calf by mail order from J. H. Miller of Peru, Indiana. Mr. Miller was recognized as one of the very best breeders of Polled Shorthorns at this time. The bull calf's name was Graceful Sultan, he was seven weeks of age, and was purchased at a cost of seven hundred dollars. Foolishly, Dad told some of his neighbors and friends about this purchase. On the day the calf was to arrive, Dad harnessed a team of mules and hitched them to a high-wheeled wagon. He took brother and me with him and we drove to Higginsville, which was about seven miles from home. The "Hummer' was due to arrive at six a.m. and was right on time. Some of Dad's neighbors and friends were also there mainly to see just how big a damn fool a fellow could be. Dad tells that the crated, hungry calf was bawling its head off and we kids started crying also, and the damn mules didn't want to back the wagon to the baggage car. No doubt it was a most trying time and one of life's darkest moments. At any rate, "Graceful Sultan" developed into a tremendous roan bull, weighing over a ton and he sired a grand group of cattle. They were cattle of great size with length of body. The cows were good milkers and they all had a docile disposition. I remember well that we exhibited at the LaFayette County Fair at Higginsville and were heavy winners.

We were getting along fine until I showed my first 4-H club steer at the American Royal. There was absolutely no culling or sifting in 1924 at the American Royal. My steer placed seventy-seventh in a class of seventy-eight. Who ever selected the stupid numbskull that judged the show must have been crazy. Needless to say, that same judge learned a hell of a lot during the next few years because we stood much higher in the classes. In later years when I have judged Junior Shows, I have made it a point to carefully look at bottom calves because I know what the youngsters are thinking.

Time went by and when I reached the advanced age of sixteen, John McRobb, who was also sixteen, and I loaded our so-called show herd in a freight car to make a circuit of several state fairs. We were blessed with having a wonderful female in our string and she went through the entire circuit as undefeated Grand Champion Female. She won in spite of the way we fed her, dressed her, or showed her. Her name was Vanity Queen and she is the only female I know of that was Grand Champion Female for three consecutive years at the International Livestock Exposition at Chicago.

We got awfully smart about fitting and showing cattle. We learned from our competitors, the judges, and particularly paid attention to the authorities from the Colleges of Agriculture and the Extension people. We used nurse cows on everything. We fed three times a day. We kept cattle in closely confined areas and in cold weather fed them cooked shelled corn and barley. We selected to win, so the shorter legged, the shorter bodied, the shorter necked, the widest and deepest bodied, and the fattest and smoothest calf is what we were looking for. We particularly tried to grow hair to make the calf appear wider and we pulled it down under the belly and flank area to give a deeper bodied appearance. When photos were taken, we dug a hole at least for the hind feet, and then built straw up to the underline. The photographer would shoot from a three-quarter angle rather than broadside to make the body appear shorter. The camera was held as high as possible to show an even shorter legged animal in the finished photo.

For many years the Shorthorn pedigree was greatly enhanced by the word "Imported". Many of our breeders went to Scotland in search of the close-to-the-ground, heavy-boned, beautiful-headed animal with an abundant hair coat. They found what they wanted and frequently with the aid of a nip or two of Scotch tea, they paid dearly for them. Many good cattle were imported, but quite a few were rather sorry specimens when they shed their Scotch hair and they didn't look so good during an American August.

The one big mistake in our efforts to glamorize our pedigrees and win in the show ring was that if a commercial man bought a bull, he possibly would be disappointed and not buy another. So, in spite of the kind that could win in the shows, some of us took a long hard look at the more practical side of beef production and bred away from the blocky, chubby little beasts. We selected and produced a far more practical kind. We wanted a calf to be of sound structure, weigh well at weaning time, weigh a thousand pounds near a year of age, top the market, and grade choice on the rail. When this kind of animal went into the show ring, it seldom won, but they gained in popularity at ringside. They were money-making cattle and this fact certainly justified our proceeding with them.

For many years about the only time a set of scales were used was when an animal was bought or sold. During the sixties for a lot of cattlemen, it became a valued asset in determining just how much a calf at seven and at twelve months weighed. As a direct result, poor producing mother cows and slow gaining cattle were beginning to be frowned on just a bit. The sonoscope was invented and it could tell us how much red meat the rib eye contained and about how much back fat covered it. Cutability of the entire animal became more important and the red meat craze was born.

The pendulum has now swung from "short" to "tall", from early maturity to late maturity, from high quality prime and overfat beef to lower quality less than choice and an all red meat animal. The belly and the brisket are gone and elevation has taken its place. To successfully compete most breeders, judges, and learned instructors will stay with the pendulum. Of course, there are still a lot of completely bewildered folks wandering around in the cattle business. There always have been a percentage of unguided or misguided souls in the purebred business and it's a reasonable assumption that there always will be. As trends change a few will be considered master breeders!

At the present time the breeder who wants to show his cattle, selects from his calf crop the longest and tallest calves he can find. The calves are put on a "growing" ration and given plenty of exercise. A few weeks before going to the show, he clips the brisket area, the entire underline and trims the sides of the belly. If he is fortunate enough to win, he will have the animal photographed. A broadside photo is taken. The calf is standing on high ground with the front leg nearest the camera just a bit forward so little or no brisket will show. The photographer squats down to give the beast all of the length of leg and elevation that is possible. We used to call this daylight. It does seem a bit stupid to see treetops just behind the topline of the animal. We are working mighty hard on this elevation characteristic because it is now in high style even though it is damned hard to cook. I often wonder just how much of it we can stand. It reminds me of the time the possum was breeding the skunk.

I would be remiss if I didn't write about the steer shows that are most entertaining and somewhat unpredictable. Years ago the ideal steer was a large fellow that would grade Prime and this meat was called "Christmas Beef". He was big, deep-bodied, long and fat. Then came the time when we didn't need big fat steers because the average size of the American family was getting smaller. It was a consensus of opinion that what we needed was a smooth, chubby, little fat steer. Baby beef was the term and early maturity was the sought for characteristic. (Today they are called quitters.) The ever-alert steer jockeys again brought out what the judges wanted. During recent years we have an entirely new trend in steer fashions. Those dumpy little fellows just are not practical cattle.

Now for the past few years the steer jockey has been guided into showing a long and tall steer with no brisket, belly, or fat. To get this greyhound type bovine, many steers have gone into extensive training with forced exercise to keep them trim. They have been fed a strictly no roughage diet so their bellies wouldn't stick out. They appear at the shows with their brisket, and underline sheared and the sides of their belly clipped in order to give this clean, no-waste, gun-barrel appearance. They are placed in the show, but finally they go the way of all steers and must be slaughtered, hung on the rail, and graded.

There's been a lot of disappointments on a great many of these carcasses. Even a lot of the Grand Champions and high placing steers have just graded good and some even go as low as standard. Any cattle feeder who makes his living in the business, knows it is impossible to trot steers every day and expect them to return a profit.

It is most interesting to me that since attending shows for over forty years and seeing all kinds of steers win, that at the wind up of each great steer show the judge is permitted or asked to speak to the crowd before he makes his final selection for Grand Champion of the show. The speech usually goes something like this: "I am honored to have been asked to pass judgment on this tremendous show of truly great cattle. In placing the classes, I have kept uppermost in my mind, every segment of the industry, including the breeder, the feeder, the packer, the retailer, and certainly most important of all, the consumer." Then he tips his hat and slaps the new Grand Champion on the rump. The sincerity of the speech and the sincerity of judgment is always entirely evident. It is most fitting for the mode of the time.

The stage is already set for the next move for the steer jockeys. They are going to stop the forced exercise. They will have to feed the calves so they will have enough finish to grade choice at one thousand or eleven hundred pounds. They will have to have a lot of red meat, but it must be well marbled meat of high quality. The entire job will have to be done economically. They will select sound calves that are correct in every way. The steer will be a fast gainer, yet when finished will have a large ribeye and excellent cutability but he will have enough bark over his carcass to grade and age properly in the cooler. He may have a limited amount of brisket, but he will also have a lot of muscling over his topline and by all means in the hind quarters. The beef from this steer will be a quality product, highly nutritious, tender, and tasty. His meat will be a delight in the fist of a small child and yet it will assert its supremacy and dignity on the finest banquet tables of presidents and kings.

The basic research for facts in this foregoing chapter are the results of over fifty years of experiments conducted on a farm at Concordia, Missouri, by a fellow who has a BS degree and had to make a living. Some of the results have been extended to you in the form of pitchfork philosophy.

▽

Dr. A. D. Weber

So much has been written and said about Dr. A. D. (Dad) Weber, his contributions so universally known and his wisdom so widely appreciated, it would take a book to summarize them all. Not only has he been one of America's greatest educators, but he is also a practical stockman's stockman as well. "Dad's" reputation is world wide, since he has judged, lectured and labored in practically every country on this planet. I've yet to hear anybody say he doesn't like Dad Weber and I doubt that anyone else has.

Dad is one of my favorite cattle judges, always retiring and reserved, but extremely capable, reasonable, efficient and thorough. Not everyone around the ringside would always agree with him, but he knew exactly what he was looking for and when he found it he could support his judgment with as plausible reasoning as anyone I've ever heard. He was about as calm and unruffled in the show ring as anyone could be even when his judgment was questioned. So far as I know he never really exploded except one time during his judgment assignments. This happened several years ago just as he had finished placing a class of steers at the International Livestock Show in Chicago. Dad had judged this show for several years and his judgment there did perhaps more than any other to right a trend by moving away from the too small unprofitable animal types.

The explosion happened like this. The steers were being judged in a ring next to where Angus breeding cattle were being placed. There was a huge crowd watching each show with some spectators alternating between the two. One was not always sure whether a spectator was watching the Angus breeding show or the Hereford steer show.

Dad finished this particular class and while he was standing chatting with Jack Turner and me, a Canadian who had been watching the Angus judging and who appeared to be unhappy with the judge's decision in the last Angus class was overheard by Dad to remark, "If that's any way to judge cattle, I'm a fool." Dad blushed firey red, grabbed the fellow, spun him around, and fairly shouted at him, "By God," he said, "I'm judging these cattle, what I say goes and if you don't like it you can get the hell out."

The poor fellow stammered an apology and almost crawled away. I'll wager he never again expressed an anti-judge opinion out in the open.

DR. ARTHUR D. WEBER

*Breed Improvement.** (Hereditary Traits in Beef Cattle.)

An infinitely wise man once said, "For there is no good tree that bringeth forth corrupt fruit; nor again a corrupt tree that bringeth forth good fruit. For of thorns men do not gather figs, nor of a bramble bush gather they grapes." And to deny that the exclusive features of purebred Herefords, Shorthorns, or Aberdeen-Angus are transmitted almost unerringly from parent to offspring would be merely to display gross ignorance of an undisputable fact.

Modern breeds of cattle arose from an almost bewildering array of colors and color combinations, types, and detailed features of conformation. Each breed emerged with an exclusive color pattern and a typical body form. Nor was the fixation of color patterns and body types accidental. Various groups of farmers had different ideas with reference to colors and types. One group preferred black, hornless cattle; a greater variety in color was more pleasing to another group, and so reds, whites, and roans were established as a breed; still other farmers were fascinated by a fixed combination of red and white, and thus the striking individuality of Herefords had its beginning.

Faith in their knowledge of heredity enabled intelligent farmers to bring a semblance of order out of chaos in cattle breeding. Broadly speaking, the rich accomplishments of the past were due to the wise selection of parental types, and the elimination of subsequent off-

Excerpts from speeches and writings of Arthur D. Weber, Vice President Emeritus, Kansas State University, and Executive Vice President, Livestock and Meat Industry Council, Inc., Manhattan, Kansas.

spring that did not measure up to a definite ideal. Vigilant selection and ruthless culling therefore have been, and still are for that matter, the tools of heredity that make for progress in animal breeding.

Widely divergent color patterns and body types are fixed sufficiently to be taken for granted. The casual observer does not notice slight variations in color or type within a breed; all Herefords, for example, look alike to him. Nevertheless, variations are still responsible for perhaps the most perplexing problem with which breeders are confronted.

Now it is true that these variations are not wholly unpredictable. It can be said with certainty that no grand champion at a major show has ever been the result of mating an inferior, almost nondescript cow with a bull of similar individuality. That it is as futile to expect cham-

pions from ancestors plainly not of championship calibre as it is to look for grapes on brambles, is almost a truism. Such pleasant surprises simply do not happen. Unpleasant surprises, however, arise from the fact that champions quite often produce offspring that do not even approach championship calibre, and that full brothers and sisters are so often unlike, irrespective of the merit of their parents.

After all, statements to the effect that we really know little or nothing of a worthwhile nature about heredity in cattle probably do not indicate intolerance or disappointment, but rather have their origin in the oft-repeated observations of seemingly unpredictable variations within established breeds. Thus, it may be that the milestones of achievement which are a silent, yet forceful tribute to man's knowledge and use in the past of the principles of heredity, are

forgotten by those who evidence anxious concern for the uncharted course that leads to future progress in cattle breeding.

The world owes a debt of gratitude to cattle breeders past and present for their noteworthy achievements. They are not discredited by the fact that the road to new and larger accomplishments that lies ahead can be charted only by the laborious efforts and careful experimentation of scientists. For it should be remembered that science can merely show how further progress may be made. Real progress will depend upon the acceptance and intelligent use of new facts by breeders themselves; hence, their future responsibilities will be far greater than those of the past.

A new approach to the solution of the now mysterious aspects of cattle inheritance came with the rediscovery of Mendelism in 1900. In obtaining new knowledge of heredity, scientists use materials and methods unfamiliar to the practical breeder. The microscope reveals hidden facts; fruit flies, plants, and small laboratory animals provide inexpensive material for experimentation; eventually newly discovered principles will be tested with large animals in breeding programs which may be confusing to practical cattlemen. Already the validity of quite a number of these new principles has been established for cattle.

Many superstitions and fears have been discarded because they are not in accord with these new principles of heredity. The idea of maternal impressions used to be quite common. For example, one of the earlier breeders of Aberdeen-Angus in Scotland kept his pregnant animals in a yard with a fence which was painted black on the inside. This was to insure the production of black calves.

Another common belief was that if a female had been bred to a particular male all her subsequent offspring would in some way resemble this particular male, even though she was mated to entirely different and unrelated males. Those who believed that this could happen were also of the opinion that if a female were continuously bred to the same male, each succeeding offspring would be more and more like the male parent. There is no foundation whatever for either of these beliefs because there is no conceivable mechanism that could bring about such effects.

Principles of heredity not only reveal why certain things cannot happen, but they also show why certain other things do happen. Inbreeding tends to intensify characteristics and make for uniformity (usually accompanied with a decline in vigor), while continued mating of cross-bred animals invariably breaks up many desirable combinations of true breeding determiners (though the cross-breds, particularly the first crosses, may be more vigorous, than their purebred ancestors). In other words, the quickest and most certain way to wipe out the results of many generations of intelligent breeding is to cross purebred animals throughout their useful breeding lives.

Of course, breeds are not sacred even though each one has a host of loyal adherents. And yet, breeds are perhaps the most important element in the livestock industry, which has grown and prospered through the years, and contributed more than any other factor to the stability of American agriculture. The great need today is not to ignore breeds or do away with them, but to strengthen each one that has proven itself; to realize that their faults should be studied and corrected if possible, not magnified and then given unwarranted publicity; in short, we should preserve that which has been gained and make certain that future accomplishments are real, not transitory. Scientists and breeders alike should direct their efforts toward the attainment of this objective.[1]

Common Cattle Can Fool You.[2]

Before me as I write this are nine bulletins and reports published from 1895 to 1933, inclusive. They give the results secured at nine agricultural experiment stations when choice or good cattle were compared with common or medium cattle. In some experiments the choice cattle were purebreds, while in others they were grades.

There is a remarkable consistency in the results obtained in these 40 years of experimentation. Unless analyzed carefully, however, the data appear to indicate that common grades of cattle, despite their inferior conformation and nondescript breeding, are able to make larger and cheaper gains when full-fed than are made by thick, beefy cattle of the better grades.

Lighter weight, thinner, common steers of nondescript breeding almost invariably made larger gains at a lower feed cost than heavier,

1. Hereditary Traits in Beef Cattle. *The Cattleman*, Vol. 23, No. 8, January, 1937. Address at Second Annual Beef Cattle Breeders Conference, Kansas State College, Manhattan, November 21, 1936.
2. Common Cattle Can Fool You—An Old Story Worth Remembering. *United States Livestock News*. Vol. 1, No. 8, November, 1934.

 fleshier beef-type steers grading good or choice. Because the common steers were poorly fed prior to the time they were placed on full-feed, they were stunted and thin. The choice steers, on the other hand, were well developed and carried more flesh.

Thin steers make larger and cheaper gains than heavier steers carrying considerable fat. This is true when common steers are compared with common steers; choice with choice, or common with choice. Grade or breeding has little to do with rate or economy of gains under such conditions. It is also well known that a thin steer makes larger gains on pasture than a fat one. That is the reason why common steers sometimes make larger gains on grass alone than choice steers.

Rate and economy of gain in the feedlot, therefore, do not provide the key to the advantage of purebred beef cattle over scrubs. The principal advantage held by choice steers over common steers is in selling price per hundredweight. If common calves are bought at a low enough price they can sell for considerably less than choice calves and make more money. But what the feeder gains the man who raised the common calves loses, so even this situation provides an argument for good, thick purebred bulls.[2]

Here Is Your Challenge, Mr. Breeder![3]

If I fail to exploit the possibilities of my subject to your satisfaction, I hope you will feel free to take up where I leave off, correct my mistakes, and add your own words of wisdom. For after all, the challenge I shall try to make is to each breeder as an individual.

At the outset, may I suggest that each of you has a personal responsibility for the progress and accomplishments of your chosen breed. You cannot delegate that responsibility to someone else. The extent to which you meet your own responsibility will be reflected in your personal success as breeder. Your breed's success represents the cumulative success of all breeders. Furthermore, the overall success of your breed serves as a ceiling, so to speak, for your own accomplishments. It is manifestly to your advantage to do all you can to encourage and help other breeders to meet their individual responsibilities. Personal success in the purebred cattle business is

3. Here is Your Challenge, Mr. Breeder! Presented at the First American Aberdeen-Angus Conference, Stillwater, Oklahoma, July 1, 1955.

never at the expense of someone else. Rather it comes only because someone else has been or will be successful. It is a sort of chain reaction, slow, to be sure, but inevitable when basic requirements have been met.

How can you best meet your challenge? What rewards should you expect, Mr. Breeder, from meeting that challenge? Past generations of breeders have sought the answers just as we today seek them and as those who follow us must, if they are to go forward. I don't have all the answers, nor do I wish to give the impression of speaking with an air of finality. All that I can hope to do is to try to arouse your interest by posing a few questions and suggesting some tentative answers. After all, it is your challenge, Mr. Breeder!

What is the challenge? In my opinion it is to make available to commercial beef herds bulls which will consistently sire more profitable calves of uniformly higher quality and of greater efficiency in converting grass, forage crops, and feed grains into beef.

I should like to emphasize strongly that the challenge is not to produce a better animal than has ever been produced before. Instead, it is a question of producing a lot of good ones, consistently and with assurance that there will be few, if any, mediocre or inferior individuals. That is the basis of profitable commercial beef production; and it is a real challenge to the breeder of purebred beef cattle to supply bulls which will consistently sire the profit-making kind without any dogie, tailenders to discredit his breeding program.

There is plenty of evidence (in 1955) that striving for the "ideal package" to the exclusion of everything else results in cattle that are too small, too fine-boned, and too close to the ground. In other words, they are an impractical and unprofitable kind from the standpoint of commercial cattlemen.

Consumers don't want to pay for bone and excess fat. Suppose that breeders decided to try to produce the type of cattle that would yield exactly the kind of carcasses that consumers appear to prefer. What sort of a beast would it be? Well, it's hard to visualize what it would be like. Some of you may recall my suggestion two or three years ago that such an animal might appropriately be dubbed a lean, muscular, boneless, fatless, gutless wonder!

Consumers' desires and preferences must be taken into account, of course, but there is manifestly a limit beyond which it is not only unprofitable but impossible to go in producing exactly what the consumer appears to want most. For it takes bone and guts and vital organs, among other things, to produce beef. Unlike manufacturers of automobiles or refrigerators or other mechanical gadgets, a cattleman markets his factory, or a replica of it, every time he makes a sale. That's what makes type complex—and it provides a real challenge to the breeder whose objective is to make available to commercial beef herds bulls which will consistently sire more profitable calves of uniformly higher quality and greater efficiency in converting grass, forage crops, and feed grains into beef.

Commercial cowmen have an intense desire to avoid fads and fancies in beef cattle. They have to keep in mind the wants and requirements of consumers, packers, grazers and feeders. They recognize the importance of the "factory" of which the brood cow is a living symbol. No one, unless it be the purebred breeder, should be more vitally concerned with strictly production factors in beef cattle than the commercial cowherd owner. And, in my opinion, commercial cowherd owners can be of inestimable help to purebred breeders who are genuinely interested in incorporating production factors into an approved type.

What are these production factors? Size is especially important—a beast with a real advantage in weight for age. Wealth of natural fleshing (muscling), great vigor and constitution, and strong bone of good quality are of prime importance. Feed handling capacity, rustling ability, and gaining ability also are very important. And milking qualities and longevity contribute immensely to usefulness of a beef cow.

Uniformity in size, color, conformation, and body type make a cowherd more appealing and assures a better market for its calf crop.

It should be reassuring to you, Mr. Breeder, that beef cattle history is replete with examples of real progress being made by those with a definite concept of what is best and then mating the best to the best. Careful selection and rigorous culling are still important means of attaining desired goals in beef cattle breeding. A keen observer, who knows animals and really likes the cattle business has a "head start," as youngsters say, in breeding better beef cattle.

Yet may I suggest, also, that there is an urgent need for new procedures and techniques for producing bulls that will consistently sire more profitable calves of uniformly higher quality and greater efficiency. All too often mating the best to the best results in too few good ones being produced. How to reduce variability in size, type, and quality of calves produced in a herd remains the number one problem in purebred beef cattle breeding.

It seems to me that you have not met your challenge fully and squarely, Mr. Breeder, if you have failed to give your whole-hearted support to the regional beef cattle breeding projects being conducted by the land-grant colleges and the United States Department of Agriculture. Now, for the first time in the history of beef cattle breeding, an organized, scientific approach is being made to the solution of your breeding problems. There is no thought of discrediting breeders or belittling their accomplishments. On the contrary, the sole objective is to be helpful, to try to find means of speeding up improvement, or reducing variability and eventually making beef cattle production more efficient and more profitable.[3]

I Tip My Hat to the Purebred Livestock Breeder.[4]

America owes the purebred livestock breeder a debt of gratitude. His leadership and steadfastness of purpose brought American agriculture to levels of excellence and efficiency unequalled in any other country. Improvement is the keystone of his operations; dignity ennobles his work; honesty and integrity characterizes his actions; and status and stability for the livestock industry are the fruits of his accomplishments.

Because of his purposeful approach to the improvement of farm animals, breeds were developed. To retain the superior qualities thus established, purebred registry associations were organized by breeders themselves. Promotional efforts of these groups extended the quality concept to all facets and phases of animal agriculture.

The influence of the purebred livestock breeder permeates all of agriculture, but consumers are the ultimate beneficiaries. Because of the continuity and effectiveness of purebred livestock breeding programs, consumers everywhere are assured of a dependable supply of nutritious meat at prices they can afford to pay. And so as one who is proud to be a meat-eating American, I tip my hat to the purebred livestock breeder!

4. I Tip My Hat to the Purebred Livestock Breeder! Beef Cattle Management Booklet. June, 1960. American Angus Association, St. Joseph, Missouri.

The Impact of Progeny and Performance Testing on the Beef Cattle Business.[5]

Emphasis in a testing program is shifted from unrecorded visual appraisals of conformation and type, and estimates of size, conformation, and weight, to birth weights, weaning weights, daily gains, etc., duly recorded for permanent record for use in breeding programs based on principles and procedures developed through research.

Progeny and performance testing is not limited to weights although, in the thinking of cattlemen generally, it has become associated with size and weight to such an extent that other aspects of the program frequently are forgotten or overlooked!

Other objective measurements can be and frequently are recorded in connection with progeny and performance testing programs. For example, body measurements, dressing percentages, cut out values, carcass weights, grades, and measurements, all have been used to a greater or lesser extent by various research workers. Likewise, calving percentage, longevity, and other traits and characteristics are being given various degrees of emphasis. Subjective appraisals of type and conformation also are being given numerical values and made a part of the performance testing records.

Yet to many breeders of purebred beef cattle, scales and the weights taken on them are the sole symbols of progeny and performance testing. To them, quality, as symbolized by type and conformation, is being overlooked, and performance records, with pounds being translated into size and rate of gain, are the primary objectives.

On the other hand, to many proponents of progeny and performance testing, including both breeders and research workers, the showring is characterized, in their thinking, by arbitrary values and uneconomic fancy points. To them, the showring symbolizes more than a century of effort to improve beef cattle by selection based on visual appraisal of type and conformation. The inevitable result, they conclude, is an impractical, outmoded show type. The only

5. The Impact of Progeny and Performance Testing on the Beef Cattle Business. August 14, 1956. A talk presented at Beef Cattle Short Course, Texas A & M College, College Station, Texas.

way to improve the situation, they insist, would be to substitute weight records in progeny and performance testing for "eye appeal" and aesthetic values.

At the risk of gaining the displeasure of both groups and being caught in the middle of the "scales versus showring" controversy, may I say that I thoroughly disagree with the viewpoints and attitudes, for and against testing, which I have just attempted to portray. It is my deep conviction that progeny and performance testing can, and in the final analysis must, combine objective measurements, weight records in particular, with subjective, and in some instances perhaps, objective, appraisals of type and conformation.

It is not a question either of using scales and liquidating beef cattle shows—or of promoting and financing bigger and bigger shows and discarding progeny and performance testing. Each has something to offer. Progeny and performance testing will have an impact on beef cattle shows, in my estimation. What its ultimate effect will be I am not prepared to say. I am convinced, however, that its influence will be beneficial and will help eventually to set forth in unmistakable terms the proper role and function of the showring in the progressive development of the beef cattle business.

To appraise what is happening and why one should try to get an over-view of the changes that have taken place in the beef cattle business during a considerable period of time and then attempt to relate those changes to economic and scientific developments.

Let's take a look first at what was happening at the turn of the century. In June, 1902, the Kansas Agricultural Experiment Station published its Bulletin No. 113 entitled "Baby Beef." Results from feeding 130 head of calves seven months are reported in the bulletin. According to the author, "The remarkable feature of this experiment is the small amount of grain required to make 100 pounds of gain."

At the end of Bulletin 113, a page of a circular is reproduced "that shows the methods of feeding and the class of cattle which were considered best forty years ago." The record is for 100 head of steers fattened by B. F. Harris, Champaign County, Illinois, and marketed at four years of age when averaging 2,377 pounds. The twelve largest cattle in the group averaged 2,786. Reference also was made to the largest steer in Illinois (in 1856), a beast seven years old weighing 3,524 pounds.

 The following comments on the last page of this bulletin are apropos of a trend which was in evidence then and has continued to the present time:

"Today the 1,000-pound steer is the standard weight to put in the feedlot, and the length of the fattening period seldom extends over seven months, and thousands of steers are in feedlots but 75 to 120 days. Other things being equal, the older the animal, the greater the amount of grain required to put on a pound of gain. The record made by Mr. Harris is published to show the strong contrast in methods and economy of feeding needed forty years ago and the methods practiced today. Is not the production of baby beef as much in advance in economy of feed and time over the usual methods of beef production of today as today's methods are over those of 1856?"

Methods of beef production of 1956 are those that were being tested in 1902. For a half century now there has been a very definite trend away from large-framed, coarse, heavy cattle. Evidences that this development would occur appeared as early as 1856, but it was not until about 1902 that the trend was clearly delineated.

Everyone with any knowledge at all of the beef cattle business has some idea of what happened and why. A steadily increasing demand for smaller cuts and the necessity for greater economy of production brought younger, lighter weight, smoother finished cattle into greater prominence on the markets. Prices reflected consumer preferences. Livestock judges responded to this development by placing more compact, thicker, blockier cattle up in the breeding classes. Packer judges, when called upon to judge steers both in carlot shows and in classes for individuals, have accentuated the trend by picking the kind of cattle they were paying the most money per pound for on the open market.

Whether market demands and prices for fat cattle on the public markets, or showring awards, were responsible for establishing the trend toward blockier, smoother beef cattle is a question concerning which there appears to be some disagreement. Really, I doubt that it matters much which should get the credit—or blame—for the striking changes in type and conformation in the major beef breeds that have occurred in the last 50 years. It is my personal feeling that type changes in the past have had their origin on the public livestock markets. Shifting needs and demands were then quickly reflected in showring awards. As a matter of fact, the sharp emphasis on type changes in the showring and quick response thereto by breeders, have tended to over-shadow the underlying causes of changing trends, in my estimation. In any event, whether we have livestock shows or not, I am convinced that market preferences, expressed in price per pound of the live animal, always have and always will help significantly to determine beef type and conformation. And I may say in passing that *medium size* is included in my concept of beef type.

There is no question but that average representatives of the major beef breeds have been changed markedly in type and conformation during the past several decades. They still vary greatly in appearance, but on the average, they are not nearly as long-legged, and rangy as formerly. But do they have as much size? Will they gain as rapidly? Are they even more profitable?

There is general agreement, I believe, that in some instances the trend toward extremely blocky, compact short-legged cattle was carried too far. This resulted perhaps from the mistaken notion in some quarters that there was a short cut to the attainment of the ideal type envisioned by most breeders.

Not all cattlemen and college animal husbandmen approved of the shift from the old to the new type beef cattle. One prominent animal husbandman persisted in placing big, off-type cattle up in the larger shows where he judged. His placings were in disfavor with the majority of exhibitors and as a consequence he no longer was invited to judge. Then there was Peter Mouw of Iowa, who was credited with having developed big-type Poland China hogs. About 40 years ago he exhibited big-type Herefords, with the idea presumably of counteracting the trend toward more compact, thicker, shorter-legged cattle. His cattle were really big, with all the characteristics usually associated with extreme size. But they proved to be unpopular and disappeared from the livestock scene.

On the other hand, there were prominent judges, college men, breeders and packer buyers, who evidenced a strong preference for the so-called typey, modern kind of cattle. Did they go too far in picking a small, pony type? A lot of people think so now, and not a few commercial cattlemen, beef cattle breeders, and college men thought so in the 1930's and 40's when this type preference was most evident in the showring.

A disturbing aspect of the situation to some purebred beef cattle breeders, however, is that, some research workers report that heritability estimates for beef type scores and conformation grades are lower than estimates for rate of gain.

Thus it seems we inevitably are brought back to that controversial subject of "scales versus showring." Would it be worthwhile for commercial cowherd owners, cattle feeders and grazers, purebred beef cattle breeders, beef breed associations, and research workers to appoint a working committee, consisting of representatives from each of these groups, to take a hard look at this question of what a beef animal should look like, and how to record differences in type, conformation, and quality so that misunderstandings and confusion could be avoided? Would such a committee reach an agreement? Or, if they failed to agree, might not their views, pro and con, when published, become a valuable contribution to beef cattle breeding literature?

Are important research opportunities in the beef cattle field being overlooked by academic administrators and research workers? And are cattlemen and those in business allied to the beef cattle industry as favorably inclined toward basic research which gives no promise whatsoever of practical discoveries as they should be? Is it true that ". . . few persons realize the frequency with which aid in the solution of practical technological problems comes from unexpected scientific directions and in ways which could not have been predicted"?

Until recently, research contributions in the field of animal science have been of much greater significance in nutrition than in breeding. There is now a growing awareness of the possibilities of applying genetic principles in the improvement of beef cattle.

When beef cattle breeders were first plagued with what appeared to be a new hereditary complex related to size and conformation (dwarfism), the challenge was met by enlisting the aid of the country's most able scientists in the fields of genetics, physiology, and animal breeding. This was an important milestone in the development of the beef cattle industry. In my judgment, the significance of this undertaking is not so much in the tangible accomplishments thus far, important as they are, but rather in the fact that a joint effort was initiated and has been recognized generally as the only sound basis for dealing with perplexing problems of this kind. Now that the pioneering has been done, we may look forward confidently to the time that many of the uncertainties of beef cattle breeding will have been eliminated through the practical application of scientific principles.

Progeny and performance testing had its origin in research findings and beef cattle breeder experiences. Its impact on the beef cattle business was accentuated by ". . . the notion, or fact–whichever was right–that our beef animals were getting a bit small . . . "

On the other hand, in my judgment, progeny and performance testing would not have been delayed long even without the apparent necessity of improving gaining ability and increasing size. For its development and application to beef cattle breeding were as logical and as inevitable as the use of hormones and antibiotics in beef cattle feeding. New procedures and practices in both beef cattle breeding and feeding are the fruits of the scientific age in which we find ourselves. We can resist changes and thus delay their adoption, or even soften their impact temporarily but in the final analysis acceptance of progeny and performance testing is the price that will have to be paid for economic survival in the beef cattle business.

And so, it is my deep conviction (in 1956) that in the years ahead progeny and performance testing will have terrific impact on the beef cattle business. I don't know when economic survival will force the majority of beef cattle breeders into progeny and performance testing–maybe in 15 or 25 years, perhaps not even then. It is to be hoped that cattlemen will not wait to adopt progeny and performance testing until economic survival forces them to do so. A real start can be made now, even though no one admittedly knows all the answers. Urgently needed research eventually will fill the gaps in existing knowledge.

The impact of progeny and performance testing on commercial herds among other things should result in better gaining cattle, and, since rate of gain tends to be correlated with economy of gain, more efficient production. There is the opportunity also to improve quality, type and conformation, but it does not necessarily follow that these improvements will automatically accompany greater gaining ability.

Progeny and performance testing also should result in higher percentage calf crops, greater lifetime production per cow, and increased profits.

The total effect of progeny and performance testing on commercial beef cattle should be a stronger competitive position through greater efficiency at each production phase—breeding, grazing, and fattening.

Similar effects doubtless will be noted in the purebred beef cattle business. As a matter of fact, as interest among commercial cowherd owners in progeny and performance testing continues to develop, there will be increasing demand for purebred bulls with performance records.

It is going to be interesting to follow progeny and performance testing with respect to the competitive aspects of the purebred beef cattle bull trade. It appears that much remains to be done in standardizing testing procedures so that records will be comparable. Environmental and other variations are complicating factors.

Interest in progeny and performance testing and the problems associated therewith should continue to stimulate beef cattle breeding research, both basic and applied, as I have indicated previously. It is to be hoped that research workers will not be expected to engage in extensive service activities in conjunction with testing programs. Their time and efforts should be used in original research, not in demonstrations. At this stage, however, many of the records in commercial and purebred herds have direct research implications or indirectly supplement formal research projects. While research data must provide the guidelines for educational programs on progeny and performance testing, it would seem that in the long run, when these programs are well established, financial support for their continuing operation should be provided by the breeder participants.

What should be the role of beef breed associations in progeny and performance testing? Breeders and their respresentatives manifestly have the responsibility of answering this question. When one takes the long view of the potential impact of progeny and performance testing on the beef cattle business, it is not inconceivable that at some future date breed associations will have assumed responsibility for technical direction and supervision of these programs for their respective breeds. There is at least a possibility, it seems to me, that the associations may add highly trained technical scientists to their staffs. It would be interesting to speculate as to the duties to which these scientists could be assigned but I shall leave that for those who would have the responsibility for directing such work.

After all, it was only in 1932 that the first proposal for progeny and performance testing of beef cattle was suggested. And it was not until 15 years later that comprehensive beef cattle breeding investigations were undertaken on a national scope. Formal recommendations and procedures for progeny testing were not forthcoming until the last few years. It is perfectly clear, therefore, that we are only at the threshold of this development. Mistakes will be made and there are bound to be disappointments.

I am confident, however, that the wisdom and good judgment of commercial cowherd owners, purebred beef cattle breeders, cattle feeders, college animal husbandmen and representatives of breed associations, the livestock press, and of allied industries will be such as to translate progeny and performance testing into a most beneficial asset to every segment of the beef cattle industry. In this far-reaching development, I predict that scales will be used extensively, and that beef cattle shows will survive. The really important consideration is, it seems to me, that we shall come to know the true possibilities and limitations of both scales and shows.[5]

Adjustment to Changes.[6]

One can scarcely pick up a newspaper or farm magazine that does not make reference to the bewildering adjustments that farmers must cope with nowadays. There are figures galore on the declining farm population; on the increasing size of farms; on trends in production per farm worker; on all sorts of things pertaining to farmers—what they do; what they earn; what they spend and owe.

The number of persons on farms has been declining for a long time and at a rapidly accelerated rate in recent years; there is an inverse relationship between the number of farmers and the size of farms; the capital outlay per farm or ranch must get larger as the enterprise increases in size; as the capital requirements for an operating unit rise, the more difficult it becomes for young men with limited resources to get established; and survival for the individual farmer or rancher depends on his having a minimum economic unit.

When an operating unit is below minimum economic size, only supplemental income from some outside source can prevent failure and eventual liquidation of the enterprise. Along with this cycle of

6. Adjustment to Changes. Address presented on Livestock Feeders' Day Program, University of Nebraska, Lincoln. April 22, 1960.

 economic consequences, the agricultural revolution has brought increased productive capacity and with it the need for greater efficiency; in other words, the ability to lower per unit production costs also has become a prime requisite for survival.

The social consequences of the agricultural revolution are far-reaching. There are those who because of age or other reasons, are unable to adjust to new situations and requirements. For them, these momentous changes do not bring progress; instead, they bring heartaches, disappointments, and disaster. Their positions are worsened; they can not hold their own; in short, retrogression is the bitter fruit of revolutionary reversals in agriculture where adjustments are difficult or impossible.

There are those, I suppose, who would halt the agricultural revolution in order to spare those who are unable to adjust to it. Perhaps they feel the damages from retrogression for the few outweigh the advantages of the agricultural revolution to the overall economy. As one who was reared on a farm and has been close to farm people all his life, I have much sympathy for those who are caught in the backwash of the momentous changes in agriculture. Certainly, there is much to be said for measures to alleviate the difficulties of those older farmers who are tied to the land and no longer possess the personal flexibility to make necessary adjustments.

Yet it is futile to try to redirect or stop the agricultural revolution in America. Actually, it is not a revolution at all, as I view it, but rather an acceleration in the last decade or so of an evolutionary process that began to pick up momentum in the United States about a 100 years ago. It started with the gradual application of scientific laws and principles to agriculture and industry. The entire economy of our country has felt the impact of this process. Indeed, it would not have been possible to have limited the application of science and technology to agriculture alone or to industry alone. The growth of one was dependent upon the simultaneous growth and development of the other. The results speak for themselves: A standard of living for all our people that is the envy of every other country in the world.

The question now is whether we shall be able to hold the gains made thus far while seeking further to extend the benefits of science and technology. I think we can. To do so, however, will require imagination of an order heretofore not attained even in America.

Furthermore, we shall have to depend even more on science and technology to chart the course of all facets of agriculture, and especially of livestock production.

Despite the far-reaching changes in American agriculture, necessary adjustments are strengthening the family-sized farm. True, it is now a commercial family farm, larger, and with more working capital than a decade ago. But it is still a family farm and will remain in that category, even though, as I pointed out earlier, there are those who because of age or other reasons find it is difficult to adjust to new situations and requirements. Too often it is not realized that of the major essentials for adjusting to changes, the human factor is perhaps the most important of all, especially in the livestock business. A stockman's job requires close attention twelve months of the year. Success comes only with intelligent study and conscientious effort. There was never any substitute for brains and hard work on a livestock farm or ranch.

Another essential for coping with changes is to keep operations in line with available resources. I knew an old-timer who used to say that you must have fat on your guts to stay in the cattle business. And a Nebraska cattle feeder had the same thing in mind when he said that the judicious cattle feeder never handles so many cattle any one year but that he can take a heavy loss and be back in business the following year.

A third essential for survival and ultimate success is a system of livestock production which is in accord with the ability of the operator, size of farm or ranch and its productive capacity, and the conservation practices in use or contemplated. Too often livestock operations represent, not a systematic program adapted to available resources and the capabilities of the owner, but an intolerable situation without purpose or plan. More general adoption of systems proved by research and experience would help greatly in adjusting to changes in the years ahead.

A fourth essential for adjusting to changes is sound financing. Capital requirements are increasing, both per farm and per farm worker. Bankers have a powerful influence on livestock operations, so it is to be hoped that they will study different livestock production systems and help to adapt these systems to the needs of individual borrowers.

Finally, there is no prospect, in my judgment, of science and technology eliminating the necessity for operative efficiency. The many details involved in the breeding, feeding, management, and marketing of livestock must be taken care of in the proper manner, at the right time, and with due regard for the influence of each one on net income.[6]

The Next 50 Years.[7]

It may appear presumptuous for anyone, and especially for a college professor, to survey the outlook for the beef cattle industry during the next 50 years. Yet the vantage point of an academician has its strong points, not the least of which is the privilege of dreaming; and dream one must if he would depict alternatives and possibilities for the beef cattle industry.

Only imaginative, practical dreamers project their thinking beyond current issues and problems. Others, all too often, are swept along by the on-rushing stream of prejudices and preconceived notions of their own generation. For them, stultifying conformity is more satisfying than creativeness. Their interest and thinking seldom are stimulated by controversial statements. They prefer the status quo, and frequently they will fight to maintain it.

Let me hasten to add, however, that recently I have noted a change in the attitudes of cattlemen, young and old. Questions are now being asked and discussions entered into with reference to shifts and adjustments in agriculture and the reasons for them. There is an eagerness now to talk about innovations that I had not detected heretofore. No longer does it seem that practical cowmen are reluctant to project their thinking into the future.

In fact, I am inclined to believe that a few rugged individualists might even be ready to admit that a little dreaming may be in order.

It is important, of course, to remember that the next 50 years will pass by one year at a time just as the last 50 years did. Nobody knows exactly what the beef cattle industry will be like 50 years hence. And I suppose a sure way to go broke would be to try to hurdle or ignore the 49 intervening years. There is no short-cut to space-age agriculture; even the most starry-eyed dreamer probably would concede that farmers and ranchers will always have to keep

7. The Next 50 Years. July 1, 1960. *The American Hereford Journal*, Vol. 51, No. 5.

their feet on the ground, figuratively speaking as well as literally.

Perhaps the strongest element in the long-term outlook for meat-animal agriculture in America is so obvious that we are inclined to take it for granted or overlook it entirely. It seems to me that each Thanksgiving we Americans should thank Divine Providence that our ancestors were meat eaters; that this great land of ours was populated by meat eaters; that we are privileged to live in a country having no taboos about meat eating; and that we can fashion our entire economy around the animal industries.

Other elements of strength in the livestock picture, both short term and in the long view, are abundant resources available for meat animal production. The grain situation not only is of tremendous concern to all cattlemen, it is also a nation-wide issue—one that will have to be resolved to restore public confidence in, and esteem for, all facets of agriculture.

How long will the United States be burdened with a surplus of grains? What is the solution to this perplexing problem?

In my judgment, two major developments eventually will relieve this troublesome situation: (1) the steadily growing human population, and (2) the effective expansion of meat animal production through maximum efficient utilization of resources, including feed grains, to provide an adequate diet for all our people, irrespective of status or income. My guess is that about 25 of the next 50 years will be required to achieve the desired interaction of these two developments. Meanwhile, the need for adjustments, with government assistance or otherwise, will continue to be debated. The situation probably will be alleviated from time to time by grain exports. It is doubtful, however, that foreign aid programs will be the final solution. On the other hand, current efforts to develop foreign outlets for grain are commendable. Assistance from every possible avenue will be needed for a long time to bring supplies in line with effective domestic and foreign demand.

The increased utilization of feed grains in producing meat for America's rapidly growing population will, I believe, be of tremendous importance in balancing our agricultural economy in the next quarter of a century. This role of meat animals tends to be overlooked just now, however, because of widespread emphasis on producing meat animals with less fat. Grain feeding is responsible for excess fat; but should less grain be utilized in livestock production?

 If so, will the growing trend toward meat-type animals aggravate rather than alleviate the surplus feed-grain situation in the years ahead?

It seems to me that in the long run increasing quantities of grain must be used in producing beef and other meat animals. The demand for carcasses from cattle weighing around 1,000 pounds and grading good to choice is almost certain to be even stronger in the years ahead. Here is one of the great challenges to beef cattle breeders and feeders: To furnish the kind of cattle and to use the feeding regimens that will supply consumers with all of the beef of the quality they want and can afford to buy. If this challenge is met during the next 50 years, and I believe it will be, then per capita beef consumption could easily reach 125 pounds. More people eating more meat—what brighter outlook could cattlemen ask for?

Long before the next half century has passed emphasis will be on stretching the grain supply, not on ways and means of disposing of surpluses. We also have vast range lands that are ideal breeding areas for beef cattle. These valuable resources can contribute to the overall economy of the country only when used for beef cattle production; and their use for this purpose will, of necessity, tend to determine the nature of complementary feeding, marketing, and merchandising enterprises throughout the entire country.

Cropping areas, both irrigated and dryland, likewise furnish tremendous quantities of plant materials that can be marketed only through livestock, and in many cases only through beef cattle.

Adjustments in cropping patterns inevitably result in new livestock producing areas. It is likely that many changes and shifts will occur in the next 50 years. Even now the South must be reckoned with in beef production. And grain-sorghum regions are now competing with the cornbelt in the fattening of cattle. In the future, few regions will enjoy distinct competitive advantages for long periods. Currently, the Northwest appears to occupy an enviable position since the demand for meat in that section of the country has outrun production, even though there are abundant resources available for livestock production. In the long run, however, this gap also will be closed and the unique competitive position will disappear.

The greatest contributions of science to livestock production thus far have been in animal nutrition and in animal pathology. Yet recent developments serve to emphasize that much more research is needed in these fields. If further advances are to be made, research emphasis should be shifted from applied research that, for example, seeks to produce more gain or to develop a new vaccine, to basic research that will tell us *why* animals gain and *how* a virus works.

I believe that during the next 50 years—yes, in the next 15 or 20 years—research in the biological sciences will be much more productive and yield results of far greater significance to mankind than will research in physical sciences. In making this prediction, I recognize, of course, that the physical sciences—chemistry, physics, and mathematics—must of necessity, be used by researchers in the field of biology. But the really significant thing is that living material, plant and animal, will be used in these studies.

Here again, I do not deprecate nor underestimate physical science research. The point is that in the past decade or so, physical science research efforts and accomplishments have been so far-reaching and significant that little attention, comparatively speaking, has been given to basic questions relating to living organisms. Now that physical science has revealed how to destroy mankind, research emphasis will, I believe, shift to biology. And it should be no more difficult to unlock the secrets of living organisms than to probe into the mysteries of the atom.

In any event, research already is underway that, I feel certain, will provide a lot of interesting and valuable information about cattle (and other farm animals), some of which might possibly be responsible for profound changes in breeding and management practices. The "how" and "why" of physiological processes will receive increasing attention. Who knows what startling developments may result from such studies? The use of new chemical compounds, hormones, etc., for example, could profoundly affect reproductive efficiency.

It is likely that research will produce a device whereby carcass characteristics can be determined objectively before slaughter. The use of ultrasonics appears promising in determining the muscling of cattle. Other procedures doubtless will be developed through research. Eventually, there will be no point in judging on-foot carcass classes because research will have provided a more accurate method of evaluation.

Research is needed to identify cattle with heavy muscling, and to establish the extent to which muscling in other parts of the carcass–forearm and rounds, for example–is associated with loin eye area. Urgently needed also are studies of relationships between beef type and carcass quality.

In a democracy, application of research results is up to the individual; but survival in a competitive economy may well depend on the extent to which the individual cattleman keeps abreast of, and uses, evolutionary developments. Yes, evolution in agriculture is gaining momentum and livestock production is beginning to feel its impact. Mechanization of feedlots is bound to come, though much experimentation will be required to evolve systems adapted to the many situations throughout the country. Whether the tremendous advances in field crop mechanization will be duplicated in livestock production remains to be seen. Drudgery will surely be removed though feedlot operations may not be fully automatic. These and many other changes will be the fruits of the evolution in livestock production. Those who ignore these developments stand to lose; indeed, attitude toward change could well determine success or failure. In the long run, those who prefer the status quo are certain to lose.

Mechanized handling of feed or automation is only one of the many intriguing changes that are bound to come. Performance testing will increase, and there will be greater emphasis on quality and carcass merit. The next 50 years will bring many changes. It would be interesting, for example, to speculate on how far specification buying of beef will be developed, what integration and contract feeding may mean to the individual producer; whether livestock shows will survive and if they do, how they must change to keep in line with other evolutionary developments in the beef cattle industry; and how breed promotion activities could and perhaps should evolve to serve the best interests of all concerned.

Of this we can be certain–imaginative men, scientists, breeders and others are going to think up and try out striking innovations. There is no reason to believe that the meat animal industry will be immune to these developments. On the contrary, "we ain't seen anything yet."

An Unpublished Postcript Dated March 25, 1972.[8]

Specialized meat animals, particularly beef cattle, are Great Britain's most notable contribution to world agriculture. Breeds were developed for efficient meat production. Dairy and draft characteristics were largely ignored, except as those traits were concomitants of reproductive and other economic objectives. The propensity to fatten was stressed, and meat quality was associated with a flavorful admixture of lean and fat in the carcass.

On the other hand, in continental Europe the objective for centuries has been to combine superior draft and milk producing characteristics in dual purpose breeds. Meat quality was never stressed, and carcasses from senile work oxen were, and still are, an important source of meat for human consumption.

British beef breeds were taken to the new world where they were the foundation of and are now the sustaining basis for the meat industry of the United States, Canada, Argentina, Australia, and New Zealand, the leading beef cattle producing countries. Continental European breeds were never a factor in the world-wide development of the livestock and meat industry; as a matter of fact, until only recently they were scarcely known outside their native districts.

Great Britain still produces substantial quantities of meat, but must import beef, pork, and lamb to meet domestic needs. Its beef imports are largely from countries to which it has supplied foundation breeding animals. This could change now that England has joined the European Common Market. The United States was once a leading supplier of beef to England, but has not figured prominently in the the meat export market for many years because of an expanded domestic market based upon population growth and increased per capita consumption.

British influence is still a potent factor in specialized meat enterprises in many countries. Yet new forces are appearing which are of great significance to American cattlemen, and, in the opinion of some authorities, presage a new era for the livestock and meat industry.

Consumer advocates are clamoring for cheaper meat, a corollary of which is pressure for larger import quotas. Changing consumer demand is being translated into breeding programs and management practices. This trend stems from growing discrimination against highly finished, including prime, beef carcasses. The campaign against "excess fat" is the outgrowth of complex interrelationships—dietary, economic, personal work habits, and faddish public re-

8. An Unpublished Postscript Dated March 25, 1972.

 sponses to propaganda. The practical effect has been a widespread search for heavier muscled, larger, faster gaining cattle, and less emphasis on early maturity and propensity to fatten.

Some now assume that the British beef breeds cannot be changed sufficiently, or in time, to meet new requirements; others apparently are not concerned one way or the other. A few may even be interested in a short-cut to the consumers pocket book, or in making a "fast buck" from sales of breeding animals, instead of developing long-term, constructive breeding programs. This is not to infer that such activities are illegal or questionable from a moral viewpoint. However, they do point up the fact that for the first time in the illustrious history of the livestock and meat industry in the United States serious attempts are being made to supplant British beef breeds with dual or triple purpose cattle (exotic breeds) from Continental Europe.

The threat to British beef breeds is not limited to the United States, as was noted in an editorial in the January, 1972, issue of the Hereford Breed Journal published in England. The writer took note of the situation by stating, among other things:

"The stream of Continental cattle flowing on to our green pastures is reawakening the cattle industry of this country. British stock is superlative and this challenge from abroad is instilling our breeders with a fresh determination to lead the world.

"The immigrants from across the English Channel have created a new interest—and they have tugged the docile British tail and stirred us into action. They are just the tonic we need and we must grasp this new chance to remind our competitors, and customers too, of the merits that have made British cattle economically superior in beef production throughout the world.

"Let it be understood that we who have made the cattle industry of this country the envy of the world, have no intention of abandoning breeding policies that are 'time tested.' All too frequently our methods are attacked and derided as being backward looking and restrictive, but our judgments are likely to be far more accurate than those of our critics, who more often than not have little or no financial involvement in breeding cattle.

"We know that no magic formula exists and that the ultimate decisions are dictated to us in the market place. At the same time we firmly believe that science and technology have a part to play and we shall always welcome sane, mature advice, not thrust upon us, but given freely in a spirit of cooperation so that we may work constructively towards the goal of keeping British breeds in the lead."

This contest comes at a time when the British beef breeds have a leading role in the most remarkable period of growth and development based on a quality concept of beef production and merchandising ever recorded in the United States. Per capita beef consumption more than doubled in the last 20 years—56 pounds in 1951 and 114 pounds in 1971. During this period the percentage of Federally graded beef designated choice climbed steadily until it now is close to 80 percent. Of the beef grading choice at least one-fourth is now making a yield grade of two or better, reflecting its high cutability. The evidence is impressive: the British beef breeds are doing a magnificent job.

Eventually, of course, the contest will be decided by comparative performances of the cattle themselves and by consumer preferences. Science and technology can help to measure performances as well as preferences, and breed associations can, and I believe will, undergird their promotional activities with scientifically verified data assembled and interpreted by their own highly qualified staff members.

Ancillary institutions and enterprises also are helping to chart the course of the livestock and meat industry in America. These include, in addition to breed associations, institutions and enterprises concerned with transportation, marketing, processing, meat merchandising, financing, science and technology, legal regulations and restrictions, news services, and commodity and general farm organizations and associations. They represent the infrastructure of America's livestock and meat industry. Their interactive relationships with production resources are so strong that nowhere in the world is there greater reason to be optimistic about the future of meat animals than in the United States.

CHAPTER 14

IT'S A PILE OF FUN

Chapter 14

It's a Pile of Fun

Every now and then certain fellows get together and predict things. Some of the things they predict are frightening.

In the so-called good old days, members of such groups were known as sages, seers or prophets. Not anymore. Today they're known as economists. Old time prophets had very few tools to work with such as computers, slide rulers and Watts line telephones. They had to rely on the Lord, crystal balls, ouija boards and corner store gossip.

Modern economists have quite a bit more going for them. They have huge stacks of charts and long shelves filled with economic guidance pamphlets, magazines, reports, statistics, weather and crop predictions and a congressman in Washington who reports back on how things are shaping up there. An ordinary fellow could get all the same material, but then an ordinary fellow wouldn't be able to understand it.

If you have in mind quitting your job and becoming an economist, maybe you'd better finish the local high school, in case you already haven't, then borrow money to take a four-year or more course in basic and advanced figures at Harvard, Cornell, Slippery Rock or Computer Tech. So far as I know, every full-fledged economist must have a Ph.D. degree. Without one he can't predict a lick and even if he could, nobody would pay him heed. I went to a school where, in those days, the emphasis was on wrestling, girl watching and stock judging with hardly anybody being concerned about economics, water, air and noise pollution or the price of long-staple cotton. Now, I'm told, all that's changed. Today the old school has a huge staff of economists—some with whiskers—the latest in computer equipment and a room full of the latest economic literature. A sneaky graduate student reveals that they still have a few ouija boards locked in a small room with a limited number of keys to same.

On the matter of farming, two favorite subjects of the economic "community" are:

1. The plight of the family farm.
2. It is pretty close to impossible for a young fellow to start in farming unless he inherits a place and money from his dad or dear old Aunt Dulcie.

It doesn't seem to be a matter of record as to exactly when these two subjects first drew the attention of the economic "community," but lately they've really warmed to the task.

To summarize the predictions, they amount to about this: the family farm is hanging on the ropes and within a short time is doomed to a knockout. Farming will be taken over by giant corporations, probably owned mostly by widows who wouldn't know a row of corn from a pit silo. The corporation will be run by a board, chairman of which is a golf playing city man, and a computer. All the plowing, seeding and harvesting will be done with huge machines, some guided by remote control from an air-conditioned office. Those who are now making a living on the family farm may as well get off their duffs and join the corporation or move to town—maybe they can get a job at a factory somewhere.

While the above is an exaggerated interpretation of the economic predictions, it is nonetheless representative of our economic thinking. Unfortunately and sadly these predictions appear to be much more than fuzzy thinking. Based on what's happened in recent years, they appear pretty close to facts. Practically nobody will question that corporate farming is on the increase. This is a fact of the times and, like it or not, we may as well accept and live with it. But increase in corporate management is not peculiar to farming, it's true of just about any business you can name.

Historically in the United States, farming and ranching have been businesses of individual initiative, generally a one-family operation with a freedom to answer to no one except Dad and the banker. Slowly but surely times are changing, even with many of the individually owned farm or ranching operations becoming family corporations. For a long time now farms have been getting larger and larger with less and less people living thereon. For example:

	Number of Farms	*Persons Living on Farms*
1940	6,096,800	30,000,000
1970	2,924,000	9,712,000

There are no secrets about why this is happening. First and likely most important is EFFICIENCY. Today, with all the modern machinery, improved seeds, feeds, fertilizers, pest controls and other scientific information, each farm worker produces enough to feed himself and 45 other people. Is it any wonder that American farm efficiency is envied by the whole world?

What about those no longer needed on the farms? Many are employed by allied agri-businesses – farm machinery manufacturing, meat and crop processing, fertilizer, insecticide and herbicide plants – and on and on. Among some who could not adapt to changes are the unemployed, hopelessly wandering the streets tempted by drugs, crime and protest.

Then there is the enormous INVESTMENT necessary to begin a successful farming or ranching operation–land, machinery, improvements, livestock, etc. It is no wonder that a young fellow panics at the prospect. Consequently, most of them hi-tail it to the city where they hope to, and usually do, get jobs. Some can't, or don't, and they, too, wind up "on the streets." Conversely, good labor each year becomes harder to get on the farm. So, farmers turn to more and better machinery. Many of those who now wander aimlessly in the ghettos could find farm or ranch employment if they were "willing to work."

Taxes on farm real estate get higher and higher each year to help support schools, roads, those on welfare relief and all the rest. Much is being advocated and done to relieve the burden of high property taxes. This is true of city and town properties as well as farm real estate. Most people, and I'm among them, believe that some way must be found to spread the tax base in order to relieve the back-breaking property levies.

And so, the corporate farming concept continues to grow. Will the family farm survive? I believe it will. In fact, I'm about to believe the peak in public corporation farming has been or soon will be reached. I hope so.

I can see several signs which support the belief that family farms and country living will not only survive but may increase in proportion to the total. Just how big is big, anyhow? No one is quite sure, but many of us ordinary people feel that corporation farms and ranches can become so large that they lose cohesiveness and that intangible thing called the personal touch. Then comes the problem of qualified labor and lack of experienced direction. Again, lest we forget, it has been individual farmers with privately owned farms plus the agri-businesses who have made American agriculture so efficient that surpluses, not scarcity, are our greatest problems.

In farming, perhaps more so than in any industry, there is that great thing called PRIDE—owning a piece of God's Own Creation. This may sound drippy, but even with certain trends to the contrary—IT IS A FACT.

In just the recent few years, we're seeing a great surge back to the country. Land is being acquired by city-employed commuters. They have a great yearning to get away from the city noises, pollution, crime in the streets and congestion of people. They want to fetch up the kids in the great outdoors with its simple pleasures and wholesome fun.

Then there's the investor, looking for a safe place to lodge a few extra dollars—perhaps he was country raised. He loves the land, so he, too, wants a slice of something of which there will be no more. Even if he continues to live in the city, you'll find him puttering in the garden and making week-end trips to smell the countryside.

If family farms are to survive and young folks attracted to them, farming MUST BE MADE ATTRACTIVE. Much has already been done. Rural electrification and water lines, good transportation, television and modern consolidated schools abound. But farmers, old and new, MUST HAVE A FAIR RETURN ON THEIR INVESTMENT. Elsewhere in this book, I've shown where, at present, the return is poor, seldom more than three percent and more often less.

Just let the price of cattle increase by a justified 10 cents per live pound, then listen to the squawks. City housewives organize boycotts, congressmen investigate and import quotas are upped.

Then about the young fellow starting the farming business. Here again the economists' predictions appear to be fairly accurate. But it CAN be done if a young man really wants to do it and provided he doesn't mind a little hard work and getting dirty. There are thousands of examples, so let me tell you about one in our community with which I am very familiar. We'll call him George—for obvious reasons not his real name.

George finished the local high school in 1964. Most of his classmates went on to college, volunteered or were drafted for military service, or went to work for someone else. Not George. He wanted to farm. But how? His father is a substantial small farmer but the home farm was too small for two farmers and there was no money to buy land and the necessary rig to set George up on his own. Besides, his parents believed it was right for a young man to hustle for himself just like they'd had to do when they started.

George sort of fooled around, as he calls it, for two years at odd jobs and helping on the home farm. Somehow he managed to save enough money to make down payments on a used truck, a little Ford tractor and a badly used combine. He was able to get credit — $2,500.00 worth—for the balance by getting his dad to sign the notes. During the two years he enrolled and finished an agricultural short course at the State University. Fortunately, George is a gifted mechanic so he could fix about anything that broke down with his battered equipment.

If he was to get under way toward becoming a farmer, he had to find ways to: (1) Fulfill his military obligation, hoping to continue his small start at least part time, (2) a source of credit which would enable him to finance his operations and (3) since he owned no land he'd have to find some he could operate on a share basis, and sufficient custom work to help pay for machinery and operation expenses.

As to the military obligations, he volunteered for the special Air Force program in which he spent week-ends at an airbase some fifty miles from home, plus two full weeks of duty in the summer right in the middle of the busiest time for farming.

On the matter of credit, because of his family's reputation, his eagerness to succeed and his willingness to work, the local banker assured him of the necessary backing.

Many small farms and a few larger ones in our area are owned by absentees or commuters, so finding sufficient share-crop land and custom work was easy.

In 1967 he bought a new "used" combine but the thing turned out to be a firetrap, having set fire to a share-crop corn field and destroying several acres of ripe corn. He swapped it in on a new one.

George continued to get more and more requests for custom and share crop work. He now was doing so well he decided it was time for the two big jumps—to get married and to start buying land of his own. Buying land came first. An eighty-acre place with a livable house on it belonging to his grandfather and his grandfather's heirs was available. So, in 1969, at a fair market price just like anyone else would have to do, George bought 39/46 of the eighty acres which was all but the 7/46 share held by his father who agreed to sell that to George when the money was available. Again his parents and he thought that he should BUY the land—not have it given to him. The price was $250 per acre which George describes as "certainly no bargain." Then he adds, "I've paid for about one-half of the eighty acres and barring a catastrophe I expect to have it all paid for in a few more years."

He married a fine local girl who works part time outside the home to help buy the groceries and furnish the modest house. He also swaps work with his father on the home farm.

How has George done financially in the five years? He showed me his financial statement which his bank accepts to secure loans for operating money.

46 cows and 44 calves, conservatively valued at	$ 20,350.00
Grain and hay on hand	7,200.00
400 acres of crops (½ of 800 acres total)	25,000.00
Two good trucks and one pick-up	10,000.00
Implements—2 tractors, combine, plows, planters, discs, etc.	20,000.00
Real estate—land and buildings	30,000.00
Cash on hand and in bank	5,000.00
Total	$117,550.00
Against this he has notes and other indebtedness	30,000.00
Net Worth	$ 87,550.00

So, here's a young man who's established a home and built up an estate worth almost a hundred thousand dollars in five short years, and he has only started. Asked if he'd do it again if he had to start all over:

"I certainly would," he said. "Sometimes it has been discouraging—eighteen long hours many and many days, bruised body and no time for boating or drag racing. But we've had a lot of fun and it's just great to see things grow and the land improve. For a little good time respite from the drudgery we enter all the local tractor-pull contests—we win most of them, too! See that hot-rod motorcycle over there—my wife and I use that for a horse. We bounce over the pasture to look after the cattle, get the work done and have fun to boot.

"We'd like to have more land," he continued, "but it's hard to find at a price that will justify farming it."

George is a fine, intelligent young man. He keeps his hair cut and washed, is clean shaven, and except for the dust and grime when he is on his job, is always neat, clean and wholesome looking. Always has a

grin from here to there. He and his young wife are an attractive couple. As long as we have their kind—and they are two of the many—the family farm will survive.

. . . AND FINALLY

—"Look What I Stepped In"—a big pile of fun ! ! !

Over forty years of meeting, working with and enjoying all the fine friends in the farming, ranching and related agri-businesses has brought great satisfaction. But the time came when putting my foot up on somebody else's gate sparked such a smoldering urge to prop my foot on my own gate that I just had to get on with it. I only hope it wasn't too late. Nothing offers greater challenge, more opportunity for creativity and a greater happiness than watching things grow--all the glorious wonders of nature.

Each morning we thank the good and gracious Lord we didn't pass on yesterday, for LOOK WHAT ANOTHER GREAT DAY WE'D HAVE MISSED ! ! !